THE CHEMISTRY OF WINE

AROMAS AND PALATES

(SECOND EDITION)

DAVID SANDUA

"Wine is a reminder that chemistry is beautiful, and that human creativity can enhance nature."

Don Kladstrup

INDEX

9

I. INTRODUCTION

The history of wine dates back thousands of years, with evidence of winemaking techniques found as early as 6000 BC in regions such as Georgia and Iran. Throughout history, wine has played a significant role in various cultures, from religious ceremonies to social gatherings. The allure of wine lies not only in its intoxicating effects but also in its ability to evoke emotions and create memorable experiences. As societies evolved, so did winemaking practices, leading to the diversity of flavors and styles seen in modern-day wines. One of the key components that contribute to the complexity of wine is its chemical makeup. Acids, sugars, tannins, and aromatic compounds all play a crucial role in determining the taste, aroma, and mouthfeel of a wine. Acids provide structure and balance, while sugars influence sweetness levels. Tannins, found primarily in red wines, add astringency and complexity, while aromatic compounds give rise to the various scents and flavors perceived when tasting a wine. Understanding these chemical components is essential for grasping why each wine possesses its unique characteristics. The fermentation process of wine is a complex interplay of yeasts and bacteria that transform grape juice into alcohol. Different strains of yeasts can produce varying flavors and aromas, while bacteria can contribute to the overall acidity of the wine. Factors such as terroir, climate, and winemaking techniques also influence the final chemical composition of a wine. By examining the chemical variations between different types of wines, from reds to whites to sparklings, a deeper appreciation for the intricate science behind winemaking can be gained.

Background of Wine Chemistry

The background of wine chemistry is rooted in the understanding of the complex interactions between different chemical compounds present in wine. The study of wine chemistry dates back to ancient times, with civilizations such as the Greeks and Romans documenting the winemaking process and experimenting with various techniques to enhance the flavor and quality of wines. Over the centuries, advancements in chemistry and technology have allowed scientists to delve deeper into the molecular composition of wine, uncovering the roles of compounds such as organic acids, sugars, and phenolic compounds in shaping the sensory characteristics of wine. This historical perspective provides a rich context for understanding the significance of wine chemistry in modern winemaking practices. The diverse chemical composition of wine is a key factor in determining its sensory properties, including color, aroma, taste, and mouthfeel. Acids, such as tartaric and malic acids, contribute to the crispness and tartness of wine, while sugars like glucose and fructose influence its sweetness levels. Tannins, derived from grape skins and seeds, provide structure and astringency to wine, affecting its texture and aging potential. Aromatic compounds, such as esters and terpenes, are responsible for the diverse range of aromas found in wine, from fruity and floral to spicy and earthy. Understanding how these chemical components interact and evolve during winemaking is crucial for producing wines of exceptional quality and character. Advancements in analytical techniques, such as chromatography and spectroscopy, have revolutionized the study of wine chemistry, allowing scientists to identify and quantify a wide range of compounds present in wine. By analyzing the chemical fingerprint of different wines, researchers can gain insights into the influence of environmental factors, such as grape variety, soil composition, and climate, on the chemical profile of wine. This knowledge is essential for winemakers seeking to produce wines that reflect the unique terroir of their vineyard and for consumers looking to appreciate the nuances of different wine styles. The background of wine chemistry encompasses a rich tapestry of scientific exploration, historical tradition, and sensory appreciation, highlighting the interdisciplinary nature of the field and its enduring relevance in the world of winemaking.

Significance of Studying Wine Chemistry

Understanding wine chemistry is of paramount significance in various fields, including agriculture, food science, and health. By studying the chemical composition of wine, researchers gain insight into the complex interactions of compounds that give rise to its distinctive flavor and aroma profiles. The presence of organic acids in wine not only influences its taste but also plays a crucial role in determining its microbial stability and preservation. The analysis of sugars in wine is essential for understanding fermentation processes and predicting the final alcohol content, sweetness, and mouthfeel of the finished product. A deeper exploration of wine chemistry provides valuable information for winemakers to enhance the quality and consistency of their products. By studying the tannins present in wine, producers can better control the astringency and aging potential of their wines, leading to improved sensory experiences for consumers. Understanding the role of aromatic compounds in wine can also help winemakers create more complex and balanced flavor profiles, attracting a wider audience and increasing market competitiveness. The study of wine chemistry empowers industry professionals to make informed decisions throughout the winemaking process, from grape selection to bottling. The study of wine chemistry has significant implications for human health and nutrition. By examining the antioxidants and polyphenols present in wine, researchers can investigate their potential health benefits, such as cardiovascular protection and anti-inflammatory properties. Understanding the chemical components of wine can also inform dietary recommendations and guidelines, promoting moderate consumption for health-conscious individuals. The exploration of wine chemistry not only enriches our understanding of this ancient beverage but also opens doors to new discoveries and applications in various scientific disciplines.

Thesis Statement

The thesis statement of this essay on the chemistry of wine asserts that understanding the chemical components of wine is essential in comprehending its unique characteristics. By exploring the intricate relationship between acids, sugars, tannins, and aromatic compounds, a deeper appreciation for the taste, aroma, and texture of wine can be gained. The chemical composition of wine is not only a result of fermentation, but also influenced by factors such as yeasts, bacteria, terroir, and winemaking techniques. Through a detailed analysis of these compounds, it becomes evident why different types of wine (red, white, rosé, sparkling) exhibit distinct flavors and traits based on their chemical makeup. The varying levels of these chemical components in wine are crucial in determining its quality, aging potential, and overall sensory experience. The balance between acidity, sweetness, bitterness, and aroma is what distinguishes a mediocre wine from an exceptional one. By delving into the roles that each compound plays in contributing to the complexity and character of wine, one can begin to appreciate the artistry and science behind winemaking. The interplay of these compounds, along with the influence of environmental factors and human intervention, adds layers of depth and nuance to the world of wine. As scientific research advances and new innovations emerge in the field of winemaking, the understanding of wine chemistry continues to evolve. Modern technologies and techniques are now being used to manipulate and enhance specific chemical compounds in order to create wines with desired attributes. By examining how these advancements are reshaping the landscape of the wine industry, it becomes clear that the intersection of chemistry and winemaking is not only fascinating but also holds great promise for the future of wine production. This journey through the compounds that create taste in wine reveals a rich tapestry of science, art, and tradition that continues to captivate enthusiasts and connoisseurs alike.

II. GRAPE COMPOSITION

The composition of grapes plays a fundamental role in determining the chemical profile of wine. Grapes contain a myriad of compounds that contribute to the taste, aroma, and texture of the final product. One of the key components in grapes are sugars, such as glucose and fructose, which are essential for the fermentation process that converts them into alcohol. Acids, including tartaric, malic, and citric acids, provide wine with its characteristic crispness and balance. Tannins, found primarily in the skins, seeds, and stems of grapes, contribute to astringency and structure in red wines, giving them their distinctive mouthfeel. In addition to sugars, acids, and tannins, grapes contain a plethora of aromatic compounds that give wine its complex bouquet. These compounds include esters, terpenes, thiols, and pyrazines, among others, each contributing unique flavors and aromas to the final product. The combination of these volatile compounds is what gives different grape varietals their distinct characteristics. The concentration of these aromatic compounds can be influenced by factors such as grape ripeness, climate, soil, and winemaking techniques, leading to a wide range of flavor profiles in the world of wine. Understanding the composition of grapes is essential for winemakers seeking to create high-quality wines. By carefully selecting grape varieties, managing vineyard practices, and controlling the winemaking process, producers can optimize the levels of sugars, acids, tannins, and aromatic compounds in their wines. This attention to detail allows winemakers to craft wines that express the unique terroir of their vineyards and showcase the full potential of the grape varieties they work with. In this way, the composition of grapes serves as the foundation for the art and science of winemaking, shaping the sensory experience that wine lovers around the world cherish.

Types of Grapes Used in Winemaking

One of the key elements in winemaking is the selection of grape varieties used in the process. Different types of grapes contribute unique flavors, aromas, and characteristics to the final product. The Cabernet Sauvignon grape is known for its bold and intense flavors, often leading to full-bodied red wines with dark fruit and tannic structure. In contrast, Chardonnay grapes are commonly used in producing white wines, known for their buttery texture and notes of apple, pear, and citrus. The choice of grape variety plays a crucial role in determining the style and profile of the wine. In addition to Cabernet Sauvignon and Chardonnay, other grape varieties are commonly used in winemaking to create a wide range of styles and flavors. Merlot grapes are favored for their soft, velvety tannins and ripe fruit characteristics, making them a popular choice for blending with other red grape varieties. Sauvignon Blanc grapes are known for their zesty and refreshing flavors, often leading to crisp and vibrant white wines with notes of green apple, pear, and grass. Each grape variety brings its own unique qualities to the winemaking process, contributing to the complexity and diversity of wines available on the market. The diversity of grape varieties used in winemaking allows for a rich tapestry of flavors and styles to be explored by wine enthusiasts. By understanding the characteristics of different grape varieties, winemakers can create wines that showcase the best of each grapes unique attributes.

Whether it be the bold flavors of Cabernet Sauvignon, the buttery texture of Chardonnay, or the zesty freshness of Sauvignon Blanc, the types of grapes chosen play a significant role in shaping the sensory experience of enjoying a glass of wine.

Chemical Components in Grapes

Grapes, the primary ingredient in winemaking, contain a rich array of chemical components that contribute to the complex flavors and aromas found in wine. One key group of compounds found in grapes are the phenolic compounds, which include tannins, anthocyanins, and flavonols. These compounds not only give wine its color and astringency, but also play a role in protecting the grape from microbial degradation. Tannins, in particular, are responsible for the dry and puckering sensation often associated with red wines, while anthocyanins provide the vibrant red, purple, or blue hues in red grape varieties. Another important chemical component in grapes is organic acids, which contribute to the overall balance and structure of wine. Malic acid, for example, gives wine its crisp and tart flavors, while tartaric acid helps stabilize the wine and prevent crystallization of potassium bitartrate. Grapes contain varying levels of sugars, mainly glucose and fructose, which are essential for the fermentation process. During fermentation, yeasts convert these sugars into alcohol, creating the alcoholic content of the wine. The delicate balance of sugars and acids in grapes is crucial in determining the final taste profile of the wine. The intricate chemical composition of grapes plays a pivotal role in shaping the sensory experience of wine. From phenolic compounds like tannins and anthocyanins, to organic acids and sugars, each component contributes to the overall flavor, aroma, and texture of the final product. Understanding the chemical makeup of grapes is essential for winemakers in crafting wines with desired characteristics. By carefully managing these chemical components during grape cultivation and winemaking processes, producers can create wines that showcase the unique terroir and craftsmanship behind each bottle.

Influence of Grape Composition on Wine Characteristics

Understanding the influence of grape composition on wine characteristics is crucial in determining the final products flavor, aroma, and mouthfeel. The grapes sugar content plays a vital role in the fermentation process, where yeast converts sugars into alcohol and carbon dioxide. Higher sugar levels result in a wine with a higher alcohol content, while lower sugar levels lead to a lighter, crisper wine. The acidity of the grapes contributes to the overall balance of the wine. Grapes with higher acidity levels produce wines with a refreshing tartness, whereas lower acidity levels can result in flabby, dull wines. In addition to sugar and acidity, the level of tannins in grapes significantly impacts the structure and mouthfeel of the wine. Tannins are polyphenolic compounds found in grape skins, seeds, and stems that create a drying sensation in the mouth. Red wines typically have higher tannin levels compared to white wines, giving them a more astringent and robust character. The presence of tannins also affects the aging potential of the wine, as they help wines develop complexity and depth over time. The aromatic compounds present in grapes contribute to the wines bouquet and flavor profile. Varietal aromas derived from grapes, such as floral, fruity, or herbal notes, can be intensified or altered during fermentation and aging, influencing the wines overall aroma. The intricate interplay of grape composition and wine characteristics underscores the importance of grape selection in winemaking. Factors such as grape variety, ripeness at harvest, and growing conditions all impact the chemical makeup of grapes and ultimately influence the final expression of the wine. By understanding how grape composition influences wine attributes, winemakers can make informed decisions during the production process to achieve the desired taste, aroma, and texture in their wines. This knowledge not only enhances the quality and complexity of the wine but also highlights the artistry and science behind winemaking.

III. FERMENTATION PROCESS

The fermentation process is a critical step in winemaking that transforms grape juice into wine through the action of yeasts. During fermentation, yeasts consume sugars in the grape juice, primarily glucose and fructose, and convert them into alcohol (ethanol) and carbon dioxide. This process, known as alcoholic fermentation, is not only responsible for the production of alcohol in wine but also influences its flavor profile. Different yeasts produce different aromatic compounds and byproducts during fermentation, which can contribute to the overall sensory experience of the wine. As yeasts metabolize sugars, they also produce primary and secondary metabolites that affect the taste and aroma of the final wine. Volatile compounds such as esters and aldehydes are produced during fermentation, adding fruity or floral notes to the wine. The presence of certain amino acids in grape juice can lead to the formation of sulfur-containing compounds, which can contribute to undesirable off-flavors if not managed properly. Controlling fermentation temperature, duration, and yeast strain selection are all critical factors that winemakers consider to optimize the fermentation process and achieve desired flavor profiles. The fermentation process in winemaking is not only a biochemical transformation but also a key factor influencing the quality and style of the final wine. The complex interactions between yeasts, sugars, and other components of grape juice during fermentation result in the unique flavor, aroma, and texture characteristics of each wine. By understanding the chemical reactions that occur during fermentation and how they impact the sensory properties of wine, winemakers can confidently navigate the winemaking process to craft wines that showcase the best of their terroir and grape varieties.

Role of Yeast in Fermentation

During the fermentation process in winemaking, yeast plays a crucial role in transforming grape sugars into alcohol, creating the desired alcoholic content of the final product. Saccharomyces cerevisiae is the most commonly used yeast strain in wine fermentation due to its ability to convert sugars into ethanol and carbon dioxide efficiently. As the yeast cells metabolize the sugars present in grape juice, they produce alcohol as a byproduct, along with other compounds that contribute to the complex flavor profile of the wine. This process of alcoholic fermentation not only affects the taste and aroma of the wine but also plays a role in preserving the final product by lowering the pH and inhibiting the growth of spoilage microorganisms. Yeast also impacts the sensory characteristics of wine through the production of volatile compounds during fermentation. These compounds, such as esters and thiols, contribute to the fruity and floral aromas that are characteristic of certain wine varieties. Yeast metabolism can also produce sulfur compounds that contribute to the overall aroma profile of the wine, adding complexity and depth to its sensory properties. Yeast strains can vary in their ability to produce certain compounds, leading to differences in flavor profiles between wines fermented with different strains. This diversity in yeast strains and their metabolic activities allows winemakers to create a wide range of wine styles with unique sensory attributes. Yeast plays a multifaceted role in the fermentation process of winemaking, impacting the chemical composition, flavor, and aroma of the final product. The ability of yeast to convert sugars into alcohol and produce a variety of volatile compounds during fermentation allows for the creation of diverse wine styles with distinct sensory characteristics. Understanding the role of yeast in fermentation is essential for winemakers to control and manipulate the chemical transformations that occur during the fermentation process, ultimately shaping the taste and quality of the wine. Further research into yeast metabolism and its interactions with other microorganisms in the winemaking process can lead to innovations in fermentation techniques and the production of wines with enhanced sensory attributes.

Chemical Reactions During Fermentation

One of the key aspects of fermentation is the complex series of chemical reactions that take place during this process. As sugars are broken down by yeast or bacteria, various compounds are produced that contribute to the flavor, aroma, and texture of the final product. During alcoholic fermentation, sugars such as glucose and fructose are converted into ethanol and carbon dioxide. This not only increases the alcohol content of the wine but also generates a range of other byproducts, including esters, phenols, and aldehydes, which add complexity to the wines profile. The type of microorganism involved in the fermentation process can greatly influence the chemical composition of the wine. Certain strains of yeast can produce higher levels of specific compounds, such as glycerol or acetic acid, which can impact the mouthfeel and acidity of the wine. The presence of lactic acid bacteria can lead to malolactic fermentation, where malic acid is converted into lactic acid, softening the acidity of the wine and introducing buttery or creamy notes. Understanding these microbial interactions is crucial for winemakers to control and manipulate the chemical profile of their wines. The chemistry of fermentation plays a crucial role in shaping the sensory characteristics of wine. By harnessing the power of yeast, bacteria, and other microorganisms, winemakers can create a diverse range of flavors and aromas in their wines. The control of temperature, oxygen exposure, and nutrient availability during fermentation can influence the balance of different compounds in the final product. Through studying the intricate chemical reactions that occur during fermentation, researchers continue to uncover new methods for enhancing the quality and uniqueness of wines around the world.

Factors Affecting Fermentation Efficiency

One key factor affecting fermentation efficiency in winemaking is the availability of nutrients for the yeast. Yeasts require essential nutrients such as nitrogen, vitamins, and minerals to carry out a successful fermentation process. A deficiency in these nutrients can lead to sluggish or stuck fermentations, resulting in lower alcohol content and poor flavor development in the wine. Winemakers must carefully monitor and adjust the nutrient levels in their must to ensure optimal conditions for yeast growth and fermentation. The presence of undesirable compounds such as pesticides or preservatives in the must can inhibit yeast activity and negatively impact fermentation efficiency. Another important factor influencing fermentation efficiency is temperature control during the fermentation process. Yeasts have specific temperature ranges at which they can operate most effectively, with different strains of yeast favoring different temperature conditions. Higher temperatures can speed up fermentation but may also lead to the production of off-flavors and aromas in the wine. On the other hand, cooler temperatures can slow down fermentation and potentially increase the risk of microbial spoilage. Winemakers must carefully manage and regulate the temperature of the fermentation vessel to create an optimal environment for yeast activity and ensure a successful fermentation with desirable wine characteristics. The type and quality of yeast used in the fermentation process can significantly impact fermentation efficiency. Different strains of yeast have varying abilities to ferment sugars, withstand alcohol levels, and produce desirable aroma compounds. Some yeast strains are known for their strong fermentation capacity under stressful conditions, while others may be more sensitive and require specific nutrient regimens. Choosing the right yeast strain for a particular wine style and grape variety is crucial for achieving the desired flavor profile and ensuring a complete and efficient fermentation. Winemakers must consider the specific attributes of the yeast strains available to them and select the most suitable option based on the desired wine characteristics and fermentation conditions.

IV. AGING AND MATURATION

Aging and maturation play crucial roles in the development of wine, contributing to its complexity and depth of flavor. As wine ages, chemical reactions continue to take place within the bottle, leading to the development of new compounds and the transformation of existing ones. One of the key processes during aging is the interaction between oxygen and the wine, which can soften harsh tannins, integrate flavors, and enhance overall balance. This oxidative process is particularly important for red wines, as it can help mellow out the sometimes aggressive tannins present in young wines, resulting in a smoother and more harmonious final product. Aging allows for the gradual breakdown of organic compounds, leading to the formation of new aromatic compounds that contribute to the wines bouquet and complexity. Maturation also plays a crucial role in the evolution of wine, influencing its texture, structure, and overall sensory profile. During maturation, compounds such as tannins and phenolic compounds undergo polymerization, leading to the development of larger molecules that can contribute to a wines mouthfeel and overall structure. This process can help soften astringent tannins, giving the wine a smoother and more rounded texture. Maturation also allows for the integration of flavors and aromas, creating a more complex and harmonious sensory experience. The duration and conditions of maturation can vary depending on the type of wine, with some wines benefiting from extended aging in oak barrels, while others are best consumed while still young and vibrant. Aging and maturation are essential components in the creation of high-quality wines, as they allow for the refinement and development of a wines flavors, aromas, and texture. Through the complex interplay of chemical reactions that take place during aging, wines can evolve and improve over time, gaining complexity, depth, and balance. Understanding the processes of aging and maturation is key for winemakers to produce wines of exceptional quality, as they must carefully consider factors such as grape variety, terroir, winemaking techniques, and aging conditions in order to create wines that showcase the best of their potential.

Oak Barrel Aging

Oak barrel aging is a crucial step in the winemaking process that significantly influences the final taste and aroma of the wine. When wine is aged in oak barrels, it undergoes a complex series of chemical reactions with the wood, resulting in the extraction of compounds that contribute to the flavor profile of the wine. The porous nature of oak allows for a gradual exchange of oxygen, which helps to soften the tannins in the wine and enhance its overall complexity. This oxidation process also imparts unique flavors and aromas to the wine, such as vanilla, caramel, and spices. Oak barrels impart tannins and other phenolic compounds to the wine, which contribute to its structure and mouthfeel. Tannins from oak are typically softer and more rounded than those from grape skins, adding a smooth and velvety texture to the wine. The interaction between the wine and the oak barrel results in the development of various esters and aldehydes, which contribute to the wines aromatic complexity. These compounds can give the wine notes of toasted oak, smoke, and nutty flavors, enhancing its overall sensory experience. Oak barrel aging plays a crucial role in shaping the flavor, aroma, and texture of wine, making it an essential component of the winemaking process. The careful selection of oak and the length of time the wine spends in the barrel can greatly impact the final product, allowing winemakers to craft wines with unique and complex characteristics. By understanding the chemical interactions that occur during oak barrel aging, winemakers can manipulate these variables to create wines that showcase the best qualities of both the grape and the wood. As a result, oak barrel aging continues to be a time-honored tradition in the world of winemaking, revered for its ability to enhance the overall quality and complexity of wines.

Chemical Changes During Aging

Recent research has highlighted the fascinating chemical changes that occur in wine as it ages, contributing to the complex flavors and aromas that develop over time. One significant chemical change during aging is the oxidation of various compounds, such as phenols and aldehydes, which can result in a mellowing of harsh flavors and a development of desirable characteristics like nuttiness or caramel notes. This process is influenced by factors like oxygen exposure, temperature, and the presence of certain enzymes, all of which can impact the overall aging trajectory of a wine. In addition to oxidation, another key chemical change during aging is the breakdown of tannins and proteins, leading to a smoother mouthfeel and enhanced complexity in the wine. Tannins, derived from grape skins and seeds, can polymerize and interact with proteins to form new compounds that contribute to the overall structure and stability of the wine. As these interactions continue over time, the tannins soften, allowing other flavors and aromas to emerge more prominently, resulting in a more harmonious and balanced wine.

The evolution of volatile compounds, such as esters and thiols, plays a crucial role in the sensory profile of aged wines. These compounds are responsible for the fruity, floral, and spicy aromas that can develop and intensify as a wine matures. Through various chemical reactions like esterification and hydrolysis, these volatile compounds can transform, interact, and evolve in a way that enhances the overall aromatic complexity of the wine, adding layers of nuance and depth that are highly prized by connoisseurs. By understanding and appreciating these chemical changes during aging, wine enthusiasts can gain a deeper insight into the intricate processes that shape the sensory experience of a fine aged wine.

Impact of Aging on Wine Flavor and Aroma

An important aspect to consider in the study of wine chemistry is the impact of aging on wine flavor and aroma. As wine matures, chemical reactions take place within the bottle that can significantly alter its sensory characteristics. One of the key changes that occur during aging is the development of complex aromas and flavors due to the breakdown of compounds like tannins and esters. Over time, the harsh and astringent components of young wines may soften, giving way to more nuanced and layered profiles that are highly sought after by wine enthusiasts. Aging can lead to the formation of new compounds that contribute to the unique bouquet of aged wines. The interactions between oxygen and phenolic compounds in the wine can lead to the formation of volatile compounds like aldehydes and ketones, which can impart desirable notes of nuttiness or caramelization. These chemical transformations not only enhance the sensory experience of the wine but also provide valuable information about its quality and potential for aging. The impact of aging on wine flavor and aroma is a complex and dynamic process that is influenced by a combination of chemical reactions. Through aging, wines can develop greater complexity, depth, and balance, making them more enjoyable and valuable to consumers. Understanding the key chemical changes that take place during aging can provide insight into how to properly age wines, as well as appreciation for the intricate science behind the development of wine flavors and aromas.

V. ACIDS AND PH LEVELS

The concentration of acids in wine directly affects its pH levels, which play a crucial role in the overall taste and quality of the beverage. Acids such as tartaric, malic, citric, and lactic acids contribute to the characteristic tartness and brightness of wine. These acids are naturally present in grapes, with their levels influenced by factors such as grape variety, climate, and vineyard practices. The pH of wine, which measures the level of acidity or alkalinity, impacts the wines stability, microbial growth, and sensory perception. A lower pH in wine is associated with higher acidity, freshness, and aging potential, while a higher pH can lead to a flabby, dull taste. Understanding the relationship between acids and pH levels is essential for winemakers to achieve the desired style and balance in their wines. Through the process of winemaking, adjustments can be made to the acidity levels by adding tartaric acid, malic acid, or citric acid to reach the desired pH range. Winemakers carefully monitor and control the pH throughout the winemaking process, from grape selection to fermentation to aging, to ensure the final product meets the expected quality standards. The precise management of acids and pH levels is a key aspect of winemaking that contributes to the complexity and character of the finished wine. The manipulation of acids and pH levels in winemaking is a delicate balancing act that requires skill and precision. The acidity of wine, determined by the presence of various acids and their concentrations, interacts with the pH to shape the sensory characteristics of the final product. By carefully adjusting and monitoring these parameters, winemakers can craft wines that exhibit balance, structure, and complexity. The study of acids and pH levels in wine not only contributes to the scientific understanding of winemaking but also enhances the appreciation of wine as a complex and multifaceted beverage.

Types of Acids Found in Wine

Acids play a crucial role in shaping the flavor profile of wine, adding complexity and balance to the overall taste. Three main types of acids are commonly found in wine: tartaric acid, malic acid, and citric acid. Tartaric acid is the most abundant acid in grapes and is responsible for the sharp, crisp taste in wines. It helps to maintain a wines pH and stability during the fermentation process. Malic acid contributes to the green apple or tart flavor in wines, and it is often found in higher concentrations in cooler climate grapes. Citric acid, while present in smaller amounts, can enhance the overall freshness and brightness of a wine, adding a subtle citrus note to the aroma. Each of these acids has a distinct impact on the sensory experience of wine. Tartaric acid, with its high acidity, can provide structure and balance to bold, full-bodied wines, while malic acid is more prevalent in lighter, crisper wines, such as many white wines. Citric acid, although in lower quantities, can contribute to the overall vibrancy and perceived freshness of a wine. The varying levels of these acids in different grape varieties and winemaking techniques contribute to the diverse array of flavors and styles found in the world of wine. Understanding the role and function of these acids can help winemakers create wines that are harmonious and well-balanced, appealing to a wide range of palates. The presence of tartaric, malic, and citric acids in wine is essential for creating a harmonious and well-rounded sensory experience. The interplay of these acids, along with other components such as sugars, tannins, and aromatic compounds, contributes to the complex flavors and textures that wine enthusiasts appreciate. By carefully managing the levels of these acids during the winemaking process, producers can achieve a desired flavor profile that reflects the characteristics of the grapes used and the region where they were grown. Understanding the types of acids found in wine is integral to appreciating the nuances and subtleties of this ancient and beloved beverage.

Importance of pH in Winemaking

pH plays a fundamental role in winemaking as it determines the overall balance and stability of the final product. By controlling the pH levels during the winemaking process, winemakers can influence crucial chemical reactions that affect the taste, aroma, and aging potential of the wine. The acidity of wine, which is directly linked to pH, is essential for preserving freshness and preventing spoilage. High acidity, achieved through maintaining a lower pH, can also enhance the perception of fruity flavors and contribute to a crisp finish in white wines. PH influences the effectiveness of sulfur dioxide, a commonly used preservative in winemaking. This compound exists in different forms depending on the pH of the wine, with the free form being the most active as an antioxidant and antimicrobial agent. By adjusting the pH, winemakers can optimize the levels of free sulfur dioxide to protect the wine from oxidation and microbial spoilage. PH impacts the stability of pigmented compounds in red wines, affecting their color intensity and longevity. Controlling the pH during fermentation and aging processes is crucial for achieving the desired hue and color stability in red wines. The importance of pH in winemaking cannot be overstated, as it influences numerous chemical reactions that shape the final characteristics of the wine. From preserving freshness and enhancing flavor profiles to ensuring microbial stability and color intensity, pH plays a central role in maintaining the quality and longevity of wines. By understanding and controlling pH levels throughout the wine-making process, producers can craft wines that resonate with consumers and showcase the unique attributes of different grape varieties and terroirs. The careful management of pH is essential for achieving consistency and excellence in winemaking.

Effects of Acid Levels on Wine Taste

Acid levels play a crucial role in determining the taste of wine, as they contribute to its overall balance and complexity. Different types of acids, such as tartaric, malic, and citric acids, can influence the perceived acidity, brightness, and crispness of a wine. When the acid levels are too low, the wine may taste flat or uninteresting, lacking the vibrancy that acidity can provide. Conversely, if the acid levels are too high, the wine can taste overly sharp or sour, overshadowing other flavors present in the wine. Finding the right balance of acidity is essential for creating a harmonious and well-rounded wine that is enjoyable to drink. Acid levels can also impact the aging potential of wine. Higher acid levels can help preserve the wine over time by acting as a natural preservative, preventing spoilage and oxidation. This is particularly important for white wines, which generally have higher acidity levels than red wines. The acidity in wine can also interact with other compounds, such as tannins, sugars, and phenolic compounds, to create a complex array of flavors and aromas. The interaction between acid and these other components can enhance the overall taste experience of the wine, making it more nuanced and interesting to the palate. Understanding the effects of acid levels on wine taste is essential for winemakers seeking to produce high-quality wines. By carefully monitoring and adjusting the acid levels during the winemaking process, they can create wines that are well-balanced, flavorful, and age-worthy. The interaction between acidity and other components in wine is a delicate dance that can greatly influence the final taste and aroma profile of the wine. As such, acidity should be viewed as a key component in the overall chemistry of wine, playing a vital role in shaping its character and quality.

VI. TANNINS AND POLYPHENOLS

Another key group of compounds found in wine are tannins and polyphenols, which are responsible for the astringency and bitterness that can be experienced when tasting certain wines. Tannins are a type of polyphenol that are primarily found in the skins, seeds, and stems of grapes, as well as in oak barrels used for aging. These compounds interact with proteins in the saliva, causing a drying and puckering sensation in the mouth. Tannins are more prevalent in red wines, especially those that are aged for longer periods, contributing to their complex structure and ability to age well. Polyphenols, on the other hand, are a broader category of compounds that include tannins as well as flavonoids and phenolic acids. These compounds are not only responsible for the bitterness and astringency in wine, but also contribute to its color, flavor, and antioxidant properties. Flavonoids, such as catechins and quercetin, are known for their health benefits, while phenolic acids like gallic acid provide a distinctive taste profile. Together, tannins and polyphenols play a crucial role in the overall composition of wine, influencing its taste, mouthfeel, and aging potential. The levels of tannins and polyphenols can vary significantly between different types of wines, with red wines typically containing higher concentrations compared to white wines. The brewing process also influences the amount and type of these compounds, with longer maceration times and oak aging leading to higher tannin levels. The interaction between tannins and polyphenols with other compounds such as acids and sugars further shapes the overall sensory experience of wine, highlighting the intricate chemistry behind its complex flavors and textures. Understanding the role of tannins and polyphenols in wine can enhance appreciation for the diverse array of flavors and characteristics found in different varietals.

Sources of Tannins in Wine

Tannins are a crucial component in wine, contributing to its structure, mouthfeel, and aging potential. These compounds are predominantly found in the skins, seeds, and stems of grapes, with red wines containing higher levels of tannins compared to white wines. Along with their astringent qualities, tannins also play a role in preserving the wine and protecting it from oxidation. The extraction of tannins during the winemaking process varies depending on the contact time between the grape solids and the juice, as well as the fermentation conditions. In addition to grape skins, seeds, and stems, oak barrels used for aging wine can also impart tannins. The toasting of oak barrels releases compounds that interact with the wine, adding complexity and enhancing its flavor profile. The choice of oak (such as French, American, or Hungarian) and the level of toast can influence the amount and type of tannins transferred to the wine. Winemakers may also use oak alternatives like chips, staves, or spirals to achieve a similar effect in a shorter period of time. Understanding the source of tannins in wine is crucial for winemakers to control and manipulate this key element in the winemaking process. The presence of tannins in wine is not limited to grape-derived sources. Some winemakers purposely add tannins extracted from other plants to enhance the wines structure and mouthfeel. These additional tannins can come from sources such as oak gall, chestnut, or quebracho, providing winemakers with a broader palette of flavors and textures to work with. The

use of non-grape tannins allows for experimentation and creativity in winemaking, leading to unique and innovative wine
styles. By understanding the various sources of tannins in wine,
winemakers can craft wines with distinctive characteristics that
stand out in the market.

Health Benefits of Polyphenols

Polyphenols, a diverse group of plant-derived compounds found
in abundance in wine, have been extensively studied for their
numerous health benefits. One of the key advantages of polyphenols is their powerful antioxidant properties. These compounds help neutralize harmful free radicals in the body, reducing oxidative stress and inflammation, which are linked to various chronic diseases. Polyphenols have been shown to have
anti-cancer properties, with some studies suggesting that they
may inhibit the growth of cancer cells and reduce the risk of
certain types of cancer. Polyphenols have been associated with
improved cardiovascular health. Resveratrol, a type of polyphenol found in red wine, has been especially well-studied for its
potential heart-protective effects. It is believed to help lower
cholesterol levels, reduce blood pressure, and improve blood
flow, all of which contribute to a healthier heart. In fact, moderate consumption of red wine has been linked to a reduced risk
of heart disease and stroke, a phenomenon often referred to as
the French Paradox. Polyphenols in wine have been shown to
support brain health and cognitive function. Some studies suggest that these compounds may help protect brain cells from
damage, improve memory and cognitive performance, and reduce the risk of neurodegenerative diseases such as Alzheimer

and Parkinson. The anti-inflammatory and antioxidant properties of polyphenols are believed to play a key role in promoting brain health and reducing the risk of age-related cognitive decline. The health benefits of polyphenols found in wine make it a valuable component of a balanced diet that promotes overall well-being.

Influence of Tannins on Wine Aging Potential

Tannins, a class of phenolic compounds found in grape skins, seeds, and stems, play a crucial role in the aging potential of wine. These compounds are responsible for the astringent and bitter qualities in wine, which can contribute to its structure and longevity. Tannins are known to interact with other wine components, such as pigmented anthocyanins and aromatic compounds, forming stable complexes that can evolve over time. This interaction can result in the softening of tannins and the development of complex flavors and aromas in aged wines. Tannins act as natural antioxidants, helping to protect wine from oxidation and microbial spoilage during the aging process. The presence of tannins in wine can help slow down the aging process, allowing the wine to develop desirable characteristics over time. The ability of tannins to bind with oxygen molecules can also contribute to the preservation of the wines color and flavor, enhancing its overall aging potential. Winemakers often carefully consider the tannin content of grapes and the extraction methods used during winemaking to optimize the aging potential of their wines. The influence of tannins on wine aging potential is a fascinating aspect of the chemistry of wine. By understanding the role of tannins in wine aging, winemakers can make informed decisions to create wines with exceptional aging potential. The complex interactions between tannins and other wine components contribute to the development of unique flavors, aromas, and textures in aged wines. The presence of tannins in wine not only enhances its structure and longevity but also contributes to the overall quality and character of the final product.

VII. SULFITES AND PRESERVATIVES

As wine is a natural product, it is susceptible to spoilage without the use of certain chemicals to preserve its quality. Sulfites are a common preservative used in winemaking to prevent oxidation and spoilage during storage and transportation. Sulfites, in the form of sulfur dioxide, also act as an antimicrobial agent, inhibiting the growth of unwanted bacteria and yeasts that could negatively impact the wine. While sulfites are essential for maintaining the stability and quality of wine, they can also have potential health risks for individuals with sulfite allergies. In addition to sulfites, winemakers may also use other preservatives such as sorbic acid and potassium sorbate to prevent microbial spoilage. These preservatives help maintain the freshness and quality of wine over time, ensuring that consumers can enjoy a consistent product with each bottle they open. While the use of preservatives in wine is necessary for ensuring its longevity and stability, there is growing consumer demand for natural and organic wines that are produced without the use of synthetic additives. The debate over the use of sulfites and other preservatives in wine highlights the tension between ensuring product quality and meeting consumer preferences for natural and organic products. Winemakers must navigate this balance carefully, considering both the technological benefits of preservatives and the consumer demand for cleaner, more natural wines. As the wine industry continues to evolve, finding innovative ways to preserve wine while meeting consumer preferences will be essential for sustainability and growth.

Role of Sulfites in Winemaking

Sulfites play a crucial role in winemaking, serving multiple functions throughout the process. One of the key roles of sulfites is as a preservative to prevent wine spoilage by inhibiting the growth of unwanted bacteria and yeasts. By releasing sulfur dioxide gas, sulfites act as a powerful antioxidant that helps to maintain the freshness and quality of the wine. This is particularly important during the aging process when wines are susceptible to oxidation and microbial contamination. Sulfites contribute to the stability of wine by preventing the formation of harmful compounds that can lead to off-flavors and aromas. Sulfites also play a role in controlling the fermentation process in winemaking. By inhibiting the activity of wild yeasts that can potentially alter the character of the wine, sulfites help winemakers to maintain control over the fermentation process. This allows for the desired flavors, aromas, and structure of the wine to be preserved, ensuring consistency in the final product. The use of sulfites during fermentation also helps to maintain the desired level of sweetness in the wine by controlling the activity of residual sugar-converting microorganisms, further shaping the overall profile of the wine. Sulfites contribute to the sensory properties of wine by influencing its color, aroma, and taste. By acting as a bleaching agent, sulfites help to preserve the natural color of the wine, preventing it from turning brown or losing its vibrancy. Sulfites also play a role in the enhancement of aromatic compounds in wine, preserving the delicate fruit and floral aromas that are characteristic of specific grape varieties. In terms of taste, sulfites help to balance acidity levels, enhancing the overall structure and mouthfeel of the wine. Thus, sulfites not only serve practical purposes in winemaking but also impact the sensory experience of the final product.

Regulations on Sulfite Use

As the demand for organic and natural products continues to rise, regulations on the use of sulfites in winemaking have become a significant topic of discussion within the industry. Sulfites, which are compounds containing sulfur dioxide, have been traditionally used in winemaking as preservatives to prevent oxidation and microbial spoilage. Concerns have been raised about the potential health risks associated with sulfite consumption, especially for individuals with sulfite sensitivity or asthma. In response to these concerns, regulatory bodies such as the Food and Drug Administration (FDA) in the United States and the European Union have implemented strict guidelines on the maximum levels of sulfites allowed in wine production. Winemakers are now faced with the challenge of finding alternative methods to preserve their wines while complying with these regulations on sulfite use. Various techniques such as using antioxidant compounds like vitamin C or natural preservatives derived from grape seeds or essential oils have been explored as alternatives to sulfites. Some winemakers have turned to more meticulous winemaking practices, such as stringent hygiene measures and temperature control, to minimize the need for sulfite use. The effectiveness and feasibility of these alternatives vary, and further research is needed to develop sustainable and cost-effective solutions for sulfite-free winemaking. The regulations on sulfite use in winemaking reflect the industries evolving efforts to balance consumer health concerns with the need for effective wine preservation. While sulfites play a crucial role in maintaining wine quality and shelf life, finding ways to reduce sulfite levels without compromising product safety remains a pressing challenge for winemakers. As advancements in technology and research continue to drive innovation in winemaking practices, the quest for sustainable and health-conscious approaches to wine preservation will undoubtedly shape the future of the industry. By embracing these challenges and exploring alternative methods to sulfite use, winemakers can not only comply with regulations but also meet the growing demand for natural and organic wines in the market.

Alternative Preservatives in Organic Wines

Preserving organic wines without compromising their quality is a challenge that winemakers face. One alternative to traditional preservatives like sulfites is the use of natural antimicrobial agents. Compounds such as lysozyme, chitosan, and potassium sorbate have been investigated for their potential to inhibit the growth of spoilage microorganisms in wine. Lysozyme, an enzyme found in egg whites, has shown promise in controlling lactic acid bacteria, which can lead to wine spoilage. Chitosan, derived from chitin in crustacean shells, has antimicrobial properties that can prevent the growth of certain wine spoilage microbes. Potassium sorbate, a salt of sorbic acid, is commonly used to prevent the growth of yeasts and molds in food and beverages, making it a potential alternative preservative for organic wines. In addition to natural antimicrobial agents, winemakers have also explored the use of essential oils and plant extracts as alternative preservatives in organic wines. Essential oils such as oregano, thyme, and cinnamon have been studied for their antimicrobial properties against wine spoilage microorganisms. These oils contain compounds like carvacrol, thymol, and cinnamaldehyde, which have been shown to inhibit the growth of bacteria and fungi. Plant extracts from grape seeds, citrus fruits, and olive leaves have also been investigated for their preservative effects in wine. These extracts contain polyphenolic compounds that possess antioxidant and antimicrobial properties, making them attractive options for preserving organic wines naturally. While the use of alternative preservatives in organic wines shows promise, more research is needed to fully understand their effectiveness and impact on wine quality. The interaction between these natural preservatives and the complex chemical composition of wine must be carefully evaluated to ensure that they do not alter the taste, aroma, or texture of the final product. By exploring new ways to preserve organic wines without the use of synthetic chemicals, winemakers can continue to produce high-quality wines that meet consumer demand for more natural and sustainable products.

VIII. WINE FAULTS AND REMEDIES

In the intricate world of winemaking, faults can arise during the fermentation, aging, or storage processes, leading to off-flavors, unpleasant aromas, or undesirable textures in the final product. One common fault is the presence of volatile acidity, caused by excessive acetic acid levels due to the action of acetic acid bacteria on ethanol. This can result in a vinegary smell and taste, which can be remedied by controlling the fermentation conditions and limiting oxygen exposure. Another fault to be aware of is cork taint, caused by trichloro anisole (TCA) contamination from cork stoppers, leading to a musty aroma and a loss of fruitiness in the wine. Preventative measures such as using alternative closures like screw caps or synthetic corks can minimize the risk of cork taint. Similarly, oxidation is a significant fault that can occur when wine is exposed to oxygen, leading to a loss of fresh fruit flavors and the development of nutty, sherry-like characteristics. To counteract oxidation, winemakers can use antioxidants such as sulfur dioxide during winemaking and bottling, as well as storing wines in a cool, dark environment to minimize oxygen exposure. Reduction, caused by a lack of oxygen during winemaking, can result in sulfurous or burnt aromas and flavors. To remedy this, winemakers can aerate the wine during production or use copper fining agents to bind with sulfur compounds and reduce their impact on the wines aroma and taste profile. Understanding the various wine faults and their remedies is crucial for producing high-quality wines that meet consumer expectations. By employing preventive measures and corrective techniques, winemakers can maintain the integrity of their wines and ensure a positive drinking experience for consumers. As the wine industry continues to evolve, advancements in technology and research will further aid winemakers in addressing faults and optimizing wine quality. Through a combination of scientific knowledge and practical experience, winemakers can navigate the complexities of wine faults and deliver exceptional products to the market.

Common Wine Faults

One of the most disappointing aspects of wine tasting is encountering common wine faults that can significantly detract from the overall experience. One of the most prevalent faults is cork taint, caused by the presence of a compound called TCA (2,4,6-trichloroanisole) which imparts a musty, moldy smell to the wine. This fault can originate from contaminated corks or wooden barrels, affecting the wines aroma and taste. Another common wine fault is oxidation, where exposure to air leads to a loss of freshness and fruity flavors in the wine. This can result from faulty closures or improper storage conditions, turning vibrant white wines into dull or brown-tinted liquids. Volatile acidity is another frequently encountered wine fault, characterized by a vinegar-like smell that masks the wines original bouquet. This fault is caused by the presence of acetic acid-producing bacteria in the winemaking process, leading to excessive levels of acetic acid. This can happen due to unsanitary conditions in the winery or lack of proper temperature control during fermentation. Microbial spoilage can occur when unwanted yeasts or bacteria contaminate the wine, resulting in off-flavors or even the formation of sediment in the bottle. These common wine faults serve as a reminder of the delicate balance needed in winemaking to ensure a high-quality, enjoyable product for consumers to savor. Understanding and identifying common wine faults is essential for both winemakers and consumers to maintain the integrity and enjoyment of wine. By recognizing the origins and characteristics of faults such as cork taint, oxidation, volatile acidity, and microbial spoilage, individuals can better appreciate the complexities of winemaking and the importance of proper storage and production practices. Through continued research and education, the wine industry can work towards reducing the occurrence of these faults and ensuring a higher standard of quality in the wines produced. Being able to pinpoint and address these issues will lead to a more pleasurable and satisfying wine tasting experience for connoisseurs and enthusiasts alike.

Chemical Causes of Wine Spoilage

One of the primary causes of wine spoilage can be attributed to chemical reactions that alter the taste, aroma, and overall quality of the wine. One common issue is the presence of acetic acid bacteria, which can lead to the formation of acetic acid, causing the wine to taste vinegary. Another culprit is ethyl acetate, produced by yeast during fermentation, which imparts a solvent-like aroma when present in high concentrations. These chemical changes can greatly diminish the sensory experience of the wine, turning what was once a pleasant drink into a disappointing one. The presence of certain compounds like hydrogen sulfide and mercaptans can also contribute to wine spoilage. These sulfur-containing compounds are often produced by yeast during fermentation under certain conditions, such as low nitrogen levels. Hydrogen sulfide can impart undesirable rotten egg or cabbage-like aromas, while mercaptans can produce foul odors reminiscent of burnt rubber. These off-putting smells can overwhelm the wines natural aromas and make it unpalatable, leading to its eventual disposal and loss of value. In addition to microbial and chemical causes, oxygen exposure is another significant factor in wine spoilage. When wines are exposed to oxygen, especially for extended periods, oxidation processes can occur, leading to the formation of aldehyde compounds that produce stale or nutty flavors. The breakdown of tannins and color compounds due to oxidation can result in browning of the wine, reducing its visual appeal. Minimizing oxygen exposure during winemaking, bottling, and storage is crucial in preserving the quality and longevity of wine.

Techniques to Correct Wine Faults

In the world of winemaking, dealing with wine faults is an inevitable challenge that requires the application of various techniques to correct. One common fault is excessive oxidation, which can lead to a wine tasting flat and dull. To address this issue, winemakers may employ techniques such as micro-oxygenation, where controlled amounts of oxygen are added to the wine to help develop its structure and enhance aromas. Another technique to correct oxidation is the addition of sulfur dioxide, a powerful antioxidant that helps preserve the wines freshness and prevent further oxidation. Blending with younger wines that have brighter fruit flavors can also help rejuvenate an oxidized wine. Another prevalent wine fault that requires correction is excessive volatile acidity, which can lead to a wine having an unpleasant vinegar-like aroma. One technique commonly used to address this issue is through the process of deacidification, where calcium carbonate or potassium bicarbonate is added to neutralize the excess acidity. Another method is to treat the wine with copper sulfate, which can help reduce volatile acidity levels. Using cold stabilization techniques can also help minimize the formation of volatile acidity compounds in the wine, preserving its balance and freshness. Another wine fault that often needs correcting is the presence of excessive sulfur compounds, which can create off-putting aromas reminiscent of burnt matches or rotten eggs. To rectify this issue, winemakers may utilize copper fining agents, such as copper sulfate or copper carbonate, which can help bind and remove sulfur compounds from the wine. Treating the wine with hydrogen peroxide can also help reduce sulfur compounds while minimizing the risk of oxidation. Allowing the wine to undergo extended aging, or even blending with wines that have lower levels of sulfur compounds, can also help mitigate these faults and improve the overall quality of the wine.

IX. WINE ANALYSIS TECHNIQUES

Various techniques are employed in the analysis of wine to understand its chemical composition and characteristics. One common method is gas chromatography, which separates the volatile compounds in wine based on their different boiling points. This allows for the identification and quantification of key aroma compounds that contribute to the overall sensory experience. Mass spectrometry is often coupled with gas chromatography to provide additional information about the molecular structure of these compounds, further enhancing our understanding of wine chemistry. These techniques are essential for quality control and ensuring the consistency of wine production. Another important analytical tool in wine analysis is liquid chromatography, which separates non-volatile compounds such as phenolic compounds, sugars, and organic acids. High-performance liquid chromatography (HPLC) is particularly useful for quantifying these components, which play a crucial role in the color, taste, and mouthfeel of wine. By analyzing the phenolic profile of wine, researchers can determine its antioxidant capacity and potential health benefits. Liquid chromatography is also used to detect any residues from pesticides or additives, ensuring that wines meet regulatory standards and are safe for consumption. In recent years, nuclear magnetic resonance (NMR) spectroscopy has emerged as a powerful technique for wine analysis. NMR can provide detailed information about the molecular structure of compounds in wine, offering insights into complex interactions between different components. By studying the chemical shifts and coupling constants in NMR spectra, researchers can identify specific compounds and monitor changes during winemaking processes. NMR spectroscopy is non-destructive and highly versatile, making it a valuable tool for studying the composition and quality of wines. These advanced analytical techniques are essential for advancing our understanding of wine chemistry and unlocking the full potential of this ancient and complex beverage.

Spectrophotometry in Wine Analysis

Spectrophotometry is a powerful analytical technique widely used in wine analysis due to its ability to quantify specific compounds present in wine. By measuring the absorption of light at various wavelengths, spectrophotometry can determine the concentration of key compounds such as phenolic compounds, sugars, and organic acids. This quantitative data provides insight into the chemical composition of wine and helps winemakers make informed decisions regarding production processes and quality control measures. Spectrophotometry can be used to assess the color intensity and hue of wines, which are important factors in evaluating appearance and potentially indicating the presence of defects. Spectrophotometric analysis plays a crucial role in evaluating the aging process of wines. Oxidation and polymerization reactions that occur over time can significantly impact the sensory properties of wine, leading to changes in color, aroma, and taste. Spectrophotometric measurements can track the evolution of phenolic compounds, which are responsible for the color and mouthfeel of wine, allowing researchers to monitor the progression of aging and assess the optimal time for bottling and consumption. Understanding these chemical changes through spectrophotometry enables winemakers to produce wines of consistent quality and character. Spectrophotometry is a valuable tool in wine analysis that provides quantitative data on key compounds, aids in quality control, and monitors the aging process of wines. By utilizing this technique, winemakers can ensure the production of high-quality wines with desirable characteristics. With its versatility and precision, spectrophotometry contributes to the advancement of wine science and the enhancement of wine production practices. As technology continues to evolve, spectrophotometric methods will likely play an increasingly important role in shaping the future of the wine industry, driving innovation and improvements in winemaking processes.

Chromatography Methods for Wine Testing

One of the key analytical techniques used in the wine industry is chromatography, a method that allows for the separation and identification of compounds present in a wine sample. Gas chromatography (GC) and liquid chromatography (LC) are commonly employed to analyze the volatile and non-volatile components of wine, respectively. GC is particularly useful for detecting aroma compounds, such as esters and alcohols, while LC is more suited for quantifying sugars, acids, and phenolic compounds. By utilizing these techniques, winemakers can gain valuable insights into the chemical composition of their wines, helping them to assess quality, detect faults, and ensure consistency. In addition to its application in quality control, chromatography methods play a vital role in wine authentication and fraud detection. As the global wine market grows, concerns about adulteration and mislabeling have become increasingly prevalent. By comparing the chromatographic profiles of a wine sample against a reference database, analysts can verify its authenticity and origins. Isotope ratio mass spectrometry (IRMS) combined with chromatography techniques can also be used to detect the addition of sugar or water, providing further assurance to consumers and regulators alike. Chromatography methods have been instrumental in research aimed at understanding the impact of winemaking practices on the chemical composition of wines. By analyzing the changes in compounds during fermentation, aging, or storage, scientists can optimize processing techniques to enhance desired characteristics and minimize undesirable traits. This knowledge can also inform decisions related to blending, oak aging, and bottle aging, allowing for the creation of wines with complex and balanced profiles. Chromatography remains a powerful tool in the quest for excellence in winemaking, offering precise and reliable analyses that underpin the art and science of wine production.

Sensory Evaluation in Wine Quality Assessment

Sensory evaluation plays a crucial role in assessing the quality of wine, providing valuable insights into its overall characteristics. This method involves the careful examination of a wines appearance, aroma, taste, and mouthfeel by trained experts or panelists. These individuals use their senses to identify specific attributes, such as fruity notes, oakiness, acidity, sweetness, and tannin levels. By systematically evaluating these sensory aspects, professionals can determine the complexity, balance, and overall quality of a wine, guiding consumer preferences and production decisions. In the middle of the assessment process, panelists pay close attention to the aroma profile of the wine, as it is a key factor in determining its quality. A wines aroma is influenced by a variety of volatile compounds, including esters, terpenes, thiols, and aldehydes, which are formed during fermentation and aging. These compounds contribute to the complex bouquet of the wine, providing clues about its grape variety, region of origin, and winemaking techniques. Panelists sniff the wine carefully, trying to identify specific aromas such as floral, fruity, spicy, or earthy notes, which can reveal important information about the wines composition and style. At the end of the evaluation, panelists focus on the taste and mouthfeel of the wine, assessing its flavor profile, acidity, sweetness, tannins, and overall balance. The taste perception is influenced by a combination of compounds, such as sugars, acids, phenolics, and alcohol, which interact on the palate to create a unique sensory experience. Panelists consider the intensity, complexity, and persistence of these flavors, as well as the wines structure and texture in the mouth. By combining the input from all sensory aspects, professionals can provide a comprehensive assessment of a wines quality, helping winemakers to improve their products and consumers to make informed choices.

X. FUTURE TRENDS IN WINE CHEMISTRY

With advancements in technology and an increasing under-standing of the chemical compounds in wine, future trends in wine chemistry are expected to revolutionize the industry. One major trend is the use of analytical techniques such as mass spectrometry and nuclear magnetic resonance spectroscopy to identify and quantify specific compounds in wine. These tools enable winemakers to better understand the impact of each compound on the final products taste and aroma, leading to more precise control over the winemaking process. Another key trend in wine chemistry is the focus on sustainability and envi-ronmentally-friendly practices. As concerns about climate change and carbon footprint continue to grow, winemakers are exploring ways to reduce chemical inputs, water usage, and en-ergy consumption in vineyard management and winemaking processes. This may involve using organic or biodynamic farm-ing methods, as well as implementing innovative technologies to minimize waste and emissions. The rise of precision wine-making, where data analytics and machine learning are used to optimize grape-growing and winemaking processes, is set to shape the future of wine production. By analyzing data on soil composition, weather patterns, grape ripeness, and fermenta-tion dynamics, winemakers can make more informed decisions to enhance the quality and consistency of their wines. These technological advancements are paving the way for a new era of wine chemistry that combines tradition with innovation to create exceptional wines for the modern consumer.

Sustainable Winemaking Practices

One of the key areas of focus in modern winemaking is the adoption of sustainable practices that minimize the environmental impact of wine production. Sustainable winemaking encompasses a range of strategies aimed at reducing the use of chemicals, water, and energy in vineyard management, grape harvesting, and wine production processes. One important aspect of sustainable winemaking is the promotion of biodiversity in vineyards, which involves planting cover crops, maintaining natural habitats for beneficial insects, and using organic fertilizers to enrich the soil. By creating a balanced ecosystem within the vineyard, sustainable winemakers are able to rely less on synthetic pesticides and fertilizers, thus preserving the health of the soil and surrounding environment. In addition to promoting biodiversity, sustainable winemaking practices also prioritize water conservation and energy efficiency. Vineyards are often located in regions where water scarcity is a concern, making efficient irrigation systems and water recycling methods crucial for sustainable production. Many wineries have implemented rainwater harvesting systems and drip irrigation technology to reduce water usage and minimize runoff. Energy consumption is another significant consideration in sustainable winemaking, with wineries increasingly investing in renewable energy sources such as solar panels and wind turbines to power their operations. By reducing their reliance on fossil fuels, wineries can significantly decrease their carbon footprint and contribute to a more environmentally friendly industry. The adoption of sustainable winemaking practices not only benefits the environment but also leads to higher quality wines with distinctive flavors and aromas. By focusing on biodiversity, water conservation, and energy efficiency, winemakers are able to cultivate healthier vines, produce grapes of exceptional quality, and craft wines that reflect the unique characteristics of the terroir. Consumers are increasingly seeking out sustainably produced wines, recognizing the efforts of winemakers to protect the environment and support local communities. As the demand for sustainable wines continues to grow, the wine industry is likely to see a shift towards more eco-friendly practices that prioritize long-term sustainability and stewardship of the land.

Nanotechnology Applications in Wine Industry

One innovative application of nanotechnology in the wine industry is the development of nano-sensors for monitoring and controlling the quality of wine during the production process. These nano-sensors can detect minute changes in the chemical composition of wine, such as acidity levels or alcohol content, in real-time, allowing winemakers to adjust conditions accordingly. By providing precise and immediate feedback, these nano-sensors help ensure consistency and quality in the final product. This technology represents a significant advancement in quality control within the industry, offering a more efficient and accurate means of monitoring wine production. Another area where nanotechnology is making a significant impact on the wine industry is in the development of nano-sized delivery systems for enhancing the sensory properties of wine. Nano-encapsulation techniques can be used to protect delicate aromatic compounds in wine, preventing their degradation and improving their release during consumption. By encapsulating these compounds in nanometer-sized structures, winemakers can enhance the aroma, flavor, and overall sensory experience of their wines. This not only contributes to the complexity and appeal of the wine but also prolongs its shelf-life and stability. Nanotechnology is playing a role in improving the packaging and preservation of wine. Nano-coatings can be applied to wine bottle surfaces to prevent oxidation and microbial contamination, extending the shelf-life of the product. Nano-materials are being used in the development of ultra-lightweight and sustainable packaging solutions for wine bottles, reducing environmental impact and transportation costs. These advancements in nanotechnology are revolutionizing the way wine is produced, packaged, and enjoyed, offering new opportunities for quality enhancement, sustainability, and consumer satisfaction in the wine industry.

Genetic Engineering for Improved Grape Varieties

Advancements in genetic engineering have opened up new possibilities for creating improved grape varieties that are better suited to specific climates, resistant to diseases, and capable of producing high-quality wines. By identifying and manipulating specific genes responsible for desired traits like disease resistance, grape scientists can develop grape varieties that require fewer chemical pesticides, reducing environmental impact and promoting sustainable wine production practices. Genetic engineering can help enhance the flavor profile of grapes by modifying genes related to sugar content, acidity levels, and aromatic compounds, ultimately leading to wines with more complex and refined characteristics. Genetic engineering can play a crucial role in preserving and protecting traditional grape varieties that are in danger of extinction due to changing environmental conditions or disease outbreaks. By introducing traits for resilience and adaptability, scientists can ensure the longevity and diversity of grapevine species, safeguarding the genetic heritage of the wine industry. Genetic engineering can speed up the breeding process for developing new grape varieties, significantly reducing the time and resources required for traditional crossbreeding methods. This accelerated approach allows researchers to respond more quickly to evolving environmental challenges and consumer preferences, ensuring a more sustainable and innovative future for grape cultivation and winemaking. Despite the potential benefits of genetic engineering for grape varieties, ethical and regulatory concerns remain significant obstacles to widespread adoption in the wine industry. Questions about the safety of genetically modified organisms (GMOs), potential impacts on biodiversity, and the long-term effects of gene editing on the environment and human health must be thoroughly addressed before genetically modified grape varieties can gain full acceptance among consumers and winemakers. As the science of genetic engineering continues to evolve, ongoing research and dialogue will be essential to navigate these complex ethical considerations and establish clear guidelines for the responsible use of genetic modification in grapevine breeding programs.

XI. WINE PACKAGING AND STORAGE

As crucial as the chemistry of wine is to its taste, the importance of proper packaging and storage cannot be overlooked. Wine packaging serves multiple functions, from protecting the wine from external factors that can degrade its quality to creating an appealing presentation for consumers. The choice of packaging material, whether glass bottles, bag-in-box, or Tetra Pak cartons, can influence the wines aging process, as different materials have varying degrees of oxygen permeability. Cork closures allow a small amount of oxygen into the bottle, aiding in the wines development, while screw caps offer a more airtight seal, preserving the wines freshness. The storage conditions of wine play a crucial role in maintaining its quality over time. Factors such as temperature, humidity, light exposure, and vibration can all impact the chemical composition of wine, leading to undesirable changes in taste and aroma. Storing wine at too high a temperature can accelerate its aging process, resulting in premature oxidation and the loss of fruitiness. Similarly, exposure to light, especially ultraviolet radiation, can cause chemical reactions in the wine that produce off-flavors and aromas. Proper storage in a cool, dark, and vibration-free environment is essential for preserving the integrity of the wines chemical compounds. Wine packaging and storage are integral aspects of maintaining the quality and character of the beverage. By understanding the chemical reactions that can occur in wine due to external factors, winemakers and consumers alike can make informed decisions about how to best protect their investment. Whether it be choosing the right packaging material to allow for controlled oxygen exposure or storing bottles in ideal conditions to prevent degradation, attention to detail in these areas can enhance the overall enjoyment of wine. The chemistry of wine extends beyond its intrinsic components to encompass the environmental factors that can shape its evolution and sensory profile over time.

Importance of Packaging in Preserving Wine Quality

One key aspect that plays a crucial role in preserving the quality of wine is the packaging in which it is stored and transported. The choice of packaging, whether it be glass bottles, corks, or alternative materials such as bag-in-box or cans, directly impacts the wines chemical composition over time. Glass bottles, for instance, provide excellent protection against oxygen and light, two factors that can lead to the degradation of wine quality. The closure used, whether it be a traditional cork or a screw cap, can influence the wines aging potential by controlling the amount of oxygen that interacts with the liquid inside. Packaging also plays a role in preventing microbial spoilage and maintaining the desired flavor profile of the wine. Oxygen exposure through faulty closures or poor packaging materials can lead to oxidation, resulting in off-flavors and aromas that detract from the wines overall quality. Proper packaging can act as a barrier against harmful bacteria and yeasts that may cause undesirable fermentation or spoilage. By ensuring a tight seal and adequate protection from external factors, the packaging can help preserve the wines freshness, aroma, and flavor compounds, allowing it to reach consumers in optimal condition. The importance of packaging in preserving wine quality cannot be overstated. It serves as a protective shield, safeguarding the delicate chemical composition of the wine from detrimental factors such as oxygen, light, and microbial contamination. By choosing the appropriate packaging materials and closures, winemakers can ensure that their products maintain their intended characteristics and flavors throughout storage and transportation. The packaging is a vital component in maintaining the integrity and quality of wine from production to consumption, contributing to a superior sensory experience for wine enthusiasts worldwide.

Types of Wine Packaging Materials

One of the key considerations in wine production is the choice of packaging material, as it not only affects the wines preservation but also its overall quality and taste. Glass bottles are the most common packaging material used for wine, as they are impermeable to oxygen and provide excellent protection against light. The dark color of wine bottles also helps to prevent oxidation, which can alter the wines flavor profile. Glass is inert, which means it does not react chemically with the wine, preserving its original characteristics. Glass bottles are heavy and can be costly to transport, leading some winemakers to explore other options. Another alternative to glass bottles is plastic packaging, which offers advantages such as being lightweight, shatterproof, and less expensive. Plastic containers are also more environmentally friendly as they can be recycled, reducing the carbon footprint of wine production. Plastic may not provide the same level of protection against oxygen and light as glass, potentially affecting the wines shelf life and flavor. Concerns have been raised about the potential for chemicals in plastic to leach into the wine, altering its taste and quality. Despite these drawbacks, plastic packaging continues to gain popularity among some winemakers seeking more sustainable options. In recent years, there has been a growing interest in alternative wine packaging materials, such as Tetra Pak cartons and bag-in-box containers. These materials offer the benefit of being lightweight, compact, and easily recyclable, reducing waste and environmental impact. Tetra Pak cartons also provide excellent protection against light and oxygen, ensuring the wines freshness and quality. Bag-in-box containers are particularly popular for everyday wines, offering convenience and extended shelf life after opening. As the wine industry continues to evolve, winemakers are likely to explore new innovative packaging materials that balance practicality, sustainability, and maintaining the integrity of the wine.

Impact of Storage Conditions on Wine Aging

One crucial aspect influencing the aging process of wine is the storage conditions in which it is kept. Temperature, humidity, light exposure, and the angle at which the bottle is stored all play a significant role in how wine evolves over time. Maintaining a consistent temperature is paramount, as fluctuations can accelerate chemical reactions within the wine, potentially compromising its quality. Ideally, wines should be stored at a temperature between 45-65 degrees Fahrenheit, with minimal fluctuations to allow for the slow, gradual aging process that enhances the flavors and aromas. In addition to temperature, the level of humidity in the storage environment can impact the aging of wine. Low humidity levels can dry out corks, leading to oxidation and potential spoilage of the wine. Conversely, high humidity can create a breeding ground for mold and mildew, which can negatively affect the wine. It is recommended to store wine in a humidity-controlled environment, typically around 70% relative humidity, to ensure the corks remain swollen and air-tight, thus preserving the wines integrity. Light exposure is another critical factor to consider when storing wine for aging. Ultraviolet light can degrade organic compounds in wine, leading to off-flavors and aromas. To protect wine from light damage, it is essential to store bottles in a dark environment or away from direct sunlight. Light-proof wine cellars or storage cabinets are ideal for long-term aging, as they provide a controlled environment that minimizes exposure to damaging light sources. By maintaining optimal storage conditions, wine enthusiasts can ensure that their wines age gracefully and develop complex flavors and aromas over time.

XII. MICROBIAL INFLUENCES ON WINE CHEMISTRY

Microbial influences play a crucial role in shaping the chemistry of wine, particularly during the fermentation process. Yeasts are responsible for converting sugars into alcohol, while bacteria can impact the acidity, flavor, and aroma of the final product. Certain yeasts, such as Saccharomyces cerevisiae, are commonly used in winemaking due to their ability to produce desirable aromatic compounds. Lactic acid bacteria can contribute to malolactic fermentation, which can soften the acidity of wine and introduce buttery aromas. The presence of specific microbes in the vineyard or winery can affect the overall chemical composition of the wine. Indigenous yeasts and bacteria present on grape skins or in the cellar environment can influence the fermentation process, leading to variations in flavor profiles. The use of wild yeasts can introduce unique characteristics to the wine, creating a sense of terroir that reflects the local environment. Understanding the role of microbial influences on wine chemistry is essential for both winemakers and consumers. By manipulating fermentation conditions or selecting specific microbial strains, winemakers can control the sensory attributes of the final product. Consumers can appreciate the diversity of flavors and aromas that arise from microbial interactions in the winemaking process. As research in this field continues to advance, novel techniques for harnessing microbial influences may further enrich the complexity and quality of wines available in the market.

Role of Microorganisms in Wine Fermentation

Microorganisms play a crucial role in wine fermentation, transforming grape juice into the complex beverage we enjoy. Yeasts, primarily Saccharomyces cerevisiae, are responsible for converting sugars into alcohol through fermentation. These microscopic organisms consume glucose and fructose in the grape juice, producing ethanol and carbon dioxide as byproducts. This fermentation process not only raises the alcohol content of the wine but also influences its flavor profile, as the metabolism of yeasts produces various compounds that contribute to the aroma and taste. In addition to yeasts, certain bacteria also participate in wine fermentation, particularly in the formation of malolactic fermentation. This secondary fermentation process, often desired in red wines, involves the conversion of malic acid into lactic acid by lactic acid bacteria. This conversion results in a softer, creamier mouthfeel and can reduce the wines overall acidity. The presence of bacteria can impact the stability of the wine and its sensory attributes. Thus, the interactions between yeasts and bacteria during fermentation are critical in determining the final chemical composition and sensory characteristics of the wine. The role of microorganisms in wine fermentation extends beyond alcohol production to shape the wines taste, aroma, and texture. These tiny agents influence not only the primary fermentation process but also secondary transformations that contribute to the complexity and balance of the final product. Understanding the interplay between yeasts, bacteria, and the grape must is essential for winemakers to achieve the desired flavor profiles and quality standards in their wines. The intricate relationships between microorganisms and wine chemistry highlight the delicate balance necessary to produce wines of exceptional character and depth.

Effects of Microbial Diversity on Wine Flavor

Microbial diversity plays a crucial role in shaping the flavor profile of wine. Throughout the fermentation process, various strains of yeasts and bacteria interact with the grape must, leading to the formation of a wide range of compounds that contribute to the complex aroma and taste of the final product. Different microbial species produce unique enzymes that can break down sugars and other molecules present in the grape juice, generating a diverse array of byproducts such as esters, aldehydes, and organic acids. These compounds can impart fruity, floral, spicy, or earthy notes to the wine, enhancing its sensory appeal and creating a distinctive flavor profile that reflects the specific microbial community involved in the fermentation process. Microbial diversity can also influence the mouthfeel and texture of wine. Certain strains of yeast and bacteria can produce polysaccharides or glycerol during fermentation, which can contribute to the body and viscosity of the wine. These compounds can enhance the perceived sweetness, smoothness, and overall mouth-coating effect of the wine, providing a more rounded and harmonious sensory experience for the consumer. By carefully selecting and managing the microbial populations present during fermentation, winemakers can control and optimize these textural attributes to craft wines with a desired mouthfeel that complements the overall flavor profile. The microbial composition of wine can evolve over time, contributing to changes in its aroma and taste as it ages. During aging, certain microorganisms present in the wine can continue to metabolize compounds, leading to further transformations and the development of complex flavor compounds. This microbial evolution can add depth and nuance to the wine, enhancing its complexity and character. Understanding the interplay between microbial diversity and wine flavor is essential for winemakers seeking to produce high-quality wines with distinctive sensory characteristics that captivate the palate of consumers and connoisseurs alike.

Strategies to Control Microbial Spoilage in Winemaking

Microbial spoilage in winemaking can pose a significant challenge to producers, as it can lead to off-flavors, aromas, and even render the wine undrinkable. One key strategy to control microbial spoilage is through the use of sulfur dioxide (SO_2), a common preservative in winemaking. SO_2 works by inhibiting the growth of spoilage microorganisms, such as acetic acid bacteria and Brettanomyces, which can produce off-flavors like vinegar or barnyard aromas. By carefully monitoring and adjusting the levels of SO_2 throughout the winemaking process, producers can effectively control microbial spoilage and ensure the quality of their wines. In addition to SO_2, another important strategy to control microbial spoilage in winemaking is through temperature control. Microorganisms thrive in certain temperature ranges, so by carefully controlling the temperature of fermentation and storage, producers can prevent the growth of spoilage microbes. Fermenting white wines at lower temperatures can help preserve delicate aromas and prevent the growth of unwanted bacteria. Similarly, storing wines at cool temperatures can slow down the activity of spoilage microorganisms and prolong the shelf life of the wine. By implementing strict temperature control measures throughout the winemaking process, producers can minimize the risk of microbial spoilage and maintain the quality of their wines. Proper sanitation and hygiene practices play a crucial role in controlling microbial spoilage in winemaking. Cross-contamination from equipment, barrels, or even the winery environment can introduce harmful microbes into the wine, leading to spoilage. By implementing rigorous sanitation protocols, such as cleaning and sanitizing equipment thoroughly between uses, using food-grade cleaning agents, and maintaining a clean winery environment, producers can minimize the risk of microbial contamination. Regular monitoring of sanitation practices and microbial populations in the winery can help identify potential issues before they impact the quality of the wine. A combination of strategies, including the use of preservatives like SO2, temperature control, and strict sanitation practices, can effectively control microbial spoilage in winemaking and ensure the production of high-quality wines.

XIII. CLIMATE AND TERROIR EFFECTS ON WINE CHEMISTRY

Climate and terroir play a crucial role in shaping the chemical composition of wine. The unique combination of temperature, precipitation, soil composition, and topography in a particular region contributes to the distinct characteristics of wines produced there. Warmer climates tend to produce wines with higher sugar levels, which can result in a higher alcohol content. On the other hand, cooler regions may produce wines with higher acidity levels, leading to a more crisp and refreshing taste. These environmental factors also influence the development of phenolic compounds, such as tannins, which contribute to the structure and aging potential of the wine. The concept of terroir encompasses not only the physical attributes of a vineyard but also the cultural, historical, and human influences that shape the final product. The interaction between the vineyard and the winemaker, as well as traditional practices passed down through generations, can impact the wine chemistry in subtle yet significant ways. The use of oak barrels during the aging process can introduce compounds like vanillin and wood tannins, adding complexity and depth to the wine. The unique terroir of a region can give rise to specific flavor profiles and aromas that reflect the essence of the land. The interplay between climate and terroir is a fundamental aspect of wine chemistry that contributes to the diversity and complexity of wines worldwide. Understanding how these factors influence the chemical composition of wine is essential for winemakers seeking to produce high-quality, distinctive wines. By paying attention to the unique terroir of their vineyards and adapting their practices to suit the climate conditions, winemakers can create wines that truly express the essence of a particular region. The study of climate and terroir effects on wine chemistry is crucial for appreciating the intricate relationship between nature, culture, and science in the art of winemaking.

Influence of Climate on Grape Ripening

One significant factor that greatly influences grape ripening is the climate in which the vines are grown. The temperature, sunlight, rainfall, and overall weather patterns in a particular region can all play a crucial role in determining when grapes reach optimal ripeness. Warmer climates tend to accelerate the ripening process, leading to higher sugar levels in the grapes, while cooler climates may result in slower ripening and lower sugar content. This difference in ripening times can directly impact the final flavor profile of the wine produced from these grapes. In regions with hot climates, the grapes are more likely to reach full ripeness, resulting in wines with higher alcohol content and riper fruit flavors. Excessive heat can also lead to overripe grapes that lack acidity, balance, and complexity in the finished wine. On the other hand, cooler climates can produce wines with more vibrant acidity, lower alcohol levels, and pronounced mineral notes due to slower ripening and longer hang time on the vine. Understanding the impact of climate on grape ripening is essential for winemakers to make informed decisions about when to harvest and how to optimize the flavor potential of their grapes. While climate is a critical factor in grape ripening, terroir also plays a significant role in shaping the final character of the wine. Terroir encompasses not only the climate but also the soil composition, topography, and vineyard altitude, all of which can influence grape development and flavor profile. By carefully managing these environmental conditions, winemakers can produce wines that showcase the unique characteristics of their specific terroir, highlighting the intricacies of the grape ripening process and creating wines with a distinct sense of place.

Terroir's Impact on Wine Characteristics

Terroirs impact on wine characteristics is a crucial aspect of winemaking that underscores the importance of the environment in which grapevines are grown. Terroir encompasses a range of factors, including soil composition, climate, topography, and altitude, all of which influence the flavor profile of the grapes and, consequently, the resulting wine. The chemical composition of the soil, for example, can affect the availability of nutrients to the vines, impacting the development of aromatic compounds and influencing the wines overall taste. Climate plays a significant role in shaping the character of wine, with the amount of sunlight, temperature fluctuations, and precipitation levels all influencing the ripening process of the grapes. Cooler climates tend to produce wines with higher acidity and more delicate flavors, while warmer climates often yield fruitier, more robust wines. Factors such as altitude and topography can affect the amount of sunlight exposure and airflow, which can further contribute to the unique characteristics of the grapes grown in a particular region. When considering the impact of terroir on wine characteristics, it is essential to recognize the intricate relationship between the environment and the chemical compounds present in the grapes. The interaction between soil, climate, and topography creates a distinct fingerprint on the grapes, which is then reflected in the final wine. Winemakers often emphasize the importance of terroir in producing wines that are true expressions of a specific region, showcasing the unique flavors and aromas that can only be achieved through the careful cultivation of grapes in a particular environment.

Chemical Changes in Grapes Due to Terroir Variations

The interaction between grapes and their environment, known as terroir, plays a crucial role in shaping the chemical composition of wine. Terroir encompasses various factors such as soil composition, climate, topography, and grape variety, all of which influence the growth and development of grapes. These variations in terroir result in distinct chemical changes in grapes, ultimately affecting the flavor, aroma, and overall quality of the wine produced. Grapes grown in a cooler climate with higher acidity levels may result in wines with a more crisp and refreshing taste, while grapes grown in warmer climates with higher sugar content may lead to richer and sweeter wines. One of the key chemical changes in grapes due to terroir variations is the accumulation of phenolic compounds, which include tannins, anthocyanins, and flavonoids. These compounds are responsible for the color, structure, and mouthfeel of wine. Different terroirs can lead to varying levels of phenolic compounds in grapes, resulting in wines with distinct color intensity, bitterness, and astringency. Terroir influences the concentration of aromatic compounds in grapes, which contribute to the wines bouquet and overall sensory experience. Factors such as soil type, irrigation practices, and altitude can all impact the production of these volatile compounds in grapes. Terroir variations can influence the levels of essential nutrients and minerals absorbed by grapevines, which in turn affect the yeast fermentation process and the production of volatile compounds in wine. Grapes grown in mineral-rich soils may result in wines with more complex flavors and aromas due to the uptake of specific nutrients by the grapevines. Understanding these chemical changes in grapes due to terroir variations is essential for winemakers to make informed decisions about vineyard management practices and to produce wines that reflect the unique characteristics of their terroir. By harnessing the potential of terroir, winemakers can create wines that not only showcase the distinctive flavors of a particular region but also exemplify the artistry and complexity of winemaking.

XIV. ENZYMES IN WINEMAKING

Enzymes play a crucial role in winemaking, serving as catalysts that facilitate various biochemical reactions during the fermentation process. One key enzyme involved in winemaking is invertase, which converts sucrose into glucose and fructose, essential sugars that are then fermented into alcohol by yeast. This enzymatic activity is vital in ensuring the proper conversion of sugars and ultimately contributes to the flavor profile of the wine. Enzymes like pectinases and cellulases are used to break down cell walls in fruit to release more juice and improve clarity in the final product. These enzymes also aid in the extraction of phenolic compounds, enhancing the color and mouthfeel of the wine. By harnessing the power of enzymes, winemakers can control and optimize various aspects of the winemaking process to achieve the desired characteristics in their wines. Enzymes are indispensable in winemaking, influencing key reactions that determine the taste, aroma, and texture of the final product. Understanding the role of enzymes in wine production allows winemakers to manipulate these biological catalysts to their advantage, fine-tuning the process to achieve the desired wine characteristics. As research advances, the use of enzymes in winemaking will likely continue to evolve, offering new possibilities for creating innovative and high-quality wines in the future.

Function of Enzymes in Wine Production

Enzymes play a crucial role in the production of wine by catalyzing key reactions that influence the flavor profile and quality of the final product. During the winemaking process, enzymes such as pectinases, cellulases, and proteases are used to break down cell walls, release sugars from grapes, and facilitate the extraction of color and flavor compounds. Pectinases, for example, help to clarify the must by breaking down pectins present in the fruit, reducing cloudiness and improving filtration efficiency. Enzymes also play a vital role in the fermentation process, where they help to convert sugars into alcohol and CO_2. Yeasts produce enzymes such as invertase and sucrase, which hydrolyze sucrose into glucose and fructose, making them available for fermentation. Enzymes produced by lactic acid bacteria can help to reduce the acidity of wine by metabolizing malic acid into lactic acid, leading to a smoother and more balanced taste. Enzymes are essential in wine production for their ability to enhance extraction, clarify the product, and improve fermentation efficiency. Enzymes are indispensable tools in the winemakers arsenal, enabling the precise control of chemical reactions that shape the characteristics of wine. By utilizing specific enzymes at different stages of the winemaking process, producers can enhance the aroma, flavor, color, and texture of their wines. Understanding the function of enzymes in wine production is essential for optimizing the quality and consistency of wines, ensuring that each bottle reflects the unique terroir and craftsmanship of the winery. As the wine industry continues to evolve, advancements in enzyme technology offer exciting opportunities for innovation and improvement in winemaking practices.

Types of Enzymes Used in Winemaking
Enzymes play a crucial role in winemaking, facilitating key reactions that ultimately shape the flavor and aroma of the final

product. Proteolytic enzymes are often used in white winemaking to break down proteins present in grape must, which can lead to a reduction in unwanted cloudiness and bitterness. These enzymes help enhance the clarity and stability of the wine, ensuring a more refined end product that is visually appealing to consumers. Pectolytic enzymes are commonly employed to aid in the extraction of color and flavor compounds from grape skins during red winemaking. By breaking down pectins that hold these compounds in place, these enzymes contribute to the rich hue and robust flavors characteristic of many red wines. Another type of enzyme widely utilized in winemaking is amylase, which helps break down starches in the grape must into fermentable sugars. This process is particularly important in ensuring a successful and efficient fermentation, as yeast rely on sugars as a source of energy to produce ethanol. By converting starches into sugars, amylase helps maintain the yeasts activity and promotes a steady fermentation process. Cellulases are enzymes that target the cellulose present in grape skins and seeds, aiding in the release of sugars and enhancing the extraction of flavors and aromas. Their action contributes to the overall complexity and depth of the wine, enriching its sensory profile and mouthfeel. Oxidative enzymes such as laccase and peroxidase play a role in managing undesirable oxidative reactions that can lead to off-flavors and aromas in wine. These enzymes help prevent browning and oxidation by breaking down phenolic compounds that are prone to oxidation, safeguarding the freshness and vibrancy of the wine. By inhibiting oxidation, these enzymes contribute to the preservation of the wines sensory characteristics and prolong its shelf life. Their protective action ensures that the wine retains its desired flavor profile and aroma bouquet, meeting the high standards expected by consumers. In employing a diverse range of enzymes, winemakers can manipulate and enhance the chemical composition of wine, shaping its overall sensory attributes and ensuring a product of exceptional quality.

Enzymatic Reactions in Wine Processing

Enzymatic reactions play a crucial role in the processing of wine, influencing its flavor, aroma, and stability. One key enzymatic reaction in wine production is the conversion of malic acid to lactic acid, known as malolactic fermentation. This process, typically carried out by lactic acid bacteria, helps reduce the acidity of wine and impart a smoother, rounder mouthfeel. Enzymes such as polyphenol oxidase and peroxidase can catalyze reactions that impact wine color and clarity by reducing the levels of phenolic compounds and preventing browning reactions. Enzymes are involved in the breakdown of complex molecules like proteins and polysaccharides during the aging of wine. Proteases and glucanases, for example, can break down proteins and polysaccharides in wine, influencing its texture and mouthfeel over time. These enzymatic reactions contribute to the development of complex flavor profiles and enhance the wines overall quality. The controlled use of enzymes in winemaking allows producers to manipulate the chemical composition of wine, producing desirable characteristics that cater to consumer preferences and market demands. Understanding the role of enzymatic reactions in wine processing is essential for producing high-quality wines with unique sensory attributes. Enzymes contribute to the complexity and characteristics of wine by influencing chemical reactions that impact flavor, aroma, color, and texture. By harnessing the power of enzymes, winemakers can fine-tune the profile of their wines, creating products that are distinct and appealing to consumers. Further research into enzymatic reactions in wine processing will continue to enhance our understanding of the chemical processes that shape the sensory experience of wine, paving the way for innovation and advancement in the wine industry.

XV. WINE ADDITIVES AND CLARIFYING AGENTS

As winemaking has evolved over time, so have the techniques and additives used to enhance the clarity and stability of the final product. Wine additives and clarifying agents play a crucial role in ensuring that the wine is visually appealing and free from any undesirable elements. Additives such as bentonite, a type of clay, are commonly used in winemaking to remove proteins and prevent haziness in white wines. Bentonite works by binding to the proteins and forming large particles that can be easily removed through filtration, resulting in a clear and crisp final product. In addition to additives, winemakers also utilize clarifying agents to further refine the appearance of the wine. Fining agents such as egg white, gelatin, and isinglass are often employed to help precipitate out suspended particles and improve the wines clarity. These agents work by attracting and binding to particles in the wine, forming larger aggregates that can settle to the bottom of the vessel, allowing for easier removal. By using a combination of additives and clarifying agents, winemakers can achieve a visually stunning product that not only tastes delicious but also appeals to the consumers aesthetic sensibilities. Despite the benefits of using additives and clarifying agents in winemaking, there is ongoing debate within the industry about the potential impact of these substances on the final product. Some argue that excessive use of additives can mask the true expression of the grapes and terroir, leading to a homogenized and less authentic wine. When used judiciously and in conjunction with traditional winemaking practices, additives and clarifying agents can be valuable tools in ensuring consistency and quality in the production of wine. It is essential for winemakers to strike a balance between innovation and tradition to create wines that are both technically sound and reflective of their unique origins.

Common Additives Used in Winemaking

One common additive used in winemaking is sulfur dioxide (SO2), which plays a crucial role in preventing oxidation and microbial spoilage. This compound acts as an antioxidant, protecting the wine from exposure to oxygen that can lead to undesirable changes in color and flavor. SO2 also has antimicrobial properties, inhibiting the growth of harmful bacteria and yeasts that could spoil the wine. Winemakers carefully regulate the amount of SO2 added to wine to ensure its effectiveness while avoiding excessive levels that could impact the taste and aroma of the final product. Sulfur dioxide is a versatile additive that helps maintain the quality and stability of wine throughout the production process. Another common additive used in winemaking is tartaric acid, which occurs naturally in grapes but is often added in small quantities to adjust the acidity of the wine. Tartaric acid contributes to the overall balance of flavors in wine, enhancing its freshness and crispness. By carefully controlling the acidity levels, winemakers can improve the structure and mouthfeel of the wine, creating a harmonious tasting experience for consumers. Tartaric acid also plays a role in stabilizing the wine, preventing the formation of crystals that can affect its appearance and quality. Tartaric acid is a valuable additive that helps fine-tune the sensory characteristics of wine while ensuring its stability over time. A third common additive used in winemaking is oak chips or barrels, which impart unique flavors and aromas to wine through the process of aging. Oak aging can add notes of vanilla, spice, and toast to the wine, enhancing its complexity and depth. The porous nature of oak allows for gradual oxidation, which can soften harsh tannins and integrate different components of the wine, resulting in a smoother and more well-rounded taste. Winemakers often choose the type of oak (such as French or American) and the level of toasting carefully to achieve the desired flavor profile for the wine. Oak is a versatile additive that can elevate the sensory experience of wine, adding layers of complexity and character that enhance its overall quality and appeal.

Role of Clarifying Agents in Wine Production

In the intricate process of winemaking, clarifying agents play a crucial role in ensuring the final product is free from unwanted particles and sediments. These agents help to clarify the wine by binding to suspended solids, proteins, and yeast cells, which can negatively impact the wines appearance, flavor, and stability. Common clarifying agents used in wine production include bentonite, gelatin, and egg whites, each with their specific properties and effectiveness in removing different types of impurities. Bentonite, a type of clay, works by attracting and binding to proteins and particles, facilitating their removal through filtration. Gelatin, derived from animal collagen, is effective in reducing tannins and color compounds, while egg whites act as a fining agent by binding to suspended particles and facilitating their precipitation. The choice of clarifying agent in winemaking is influenced by various factors, including the type of wine being produced, the desired level of clarity, and the winemakers preferences. Some agents are better suited for red wines, where tannins and color compounds are more prevalent, while others are more effective in clarifying white wines, which may contain higher levels of proteins. The timing of adding clarifying agents during the winemaking process can also impact their effectiveness. Bentonite is often added early in the fermentation process to bind to proteins and promote the settling of solids, while gelatin and egg whites are typically added later in the process to fine-tune the wines clarity and stability. The use of clarifying agents in wine production not only helps improve the visual appeal of the final product but also plays a significant role in enhancing its flavor profile and stability. By effectively removing unwanted particles and impurities, clarifying agents contribute to a cleaner, more refined wine that is free from off-flavors and sedimentation. Winemakers must carefully consider the choice and timing of clarifying agents to ensure that they achieve the desired level of clarity and quality in their wines, highlighting the importance of these agents in the complex chemistry of winemaking.

Effects of Additives on Wine Stability and Quality

One key aspect that can greatly influence the stability and quality of wine is the use of additives during the winemaking process. Additives encompass a wide range of compounds that are utilized to modify or enhance specific characteristics of wine, such as color, flavor, aroma, and stability. Sulfur dioxide (SO_2) is a common additive used in winemaking to prevent oxidation and microbial spoilage, thereby preserving the wines freshness and shelf-life. Fining agents such as bentonite or egg whites are used to clarify wine by removing unwanted particles that can cause cloudiness. The effects of additives on wine stability and quality can vary depending on the type and concentration of the additive used. Excessive use of SO2 can lead to undesirable sensory characteristics in wine, such as a pronounced sulfur aroma or taste. On the other hand, the proper application of fining agents can result in a more visually appealing and clear wine, enhancing its overall quality. It is essential for winemakers to carefully consider the potential impact of additives on wine stability and quality, striking a balance between achieving desired outcomes and avoiding negative consequences. The judicious use of additives in winemaking plays a crucial role in shaping the stability and quality of the final product. By understanding the specific functions and effects of different additives, winemakers can make informed decisions throughout the winemaking process to achieve their desired outcomes. Through proper application and monitoring of additives, winemakers can ensure that the wine maintains its integrity, flavor profile, and overall quality, meeting the expectations of consumers and industry standards. As such, the study and application of additives in winemaking continues to be a vital area of research and innovation in the wine industry.

XVI. BIOGENIC AMINES IN WINE

Biogenic amines are naturally occurring compounds found in many foods, including wine. These compounds are produced during fermentation by the metabolism of amino acids by yeast and bacteria. In wine, the most common biogenic amines include histamine, tyramine, and putrescine, among others. While these compounds are typically present in low concentrations, they can have significant implications for individuals who are sensitive to them. Histamine, for example, can cause allergic reactions in some people, while tyramine has been linked to migraines in certain individuals. Putrescine, on the other hand, has a foul odor and may contribute to off-flavors in wine if present in high amounts. The presence of biogenic amines in wine is influenced by various factors, including the grape varieties used, fermentation conditions, and the presence of specific microorganisms. Certain yeast strains are known to produce higher levels of biogenic amines compared to others. The use of sulfites in winemaking can affect the production of biogenic amines, as these compounds can inhibit the growth of amine-producing bacteria. Winemakers must consider these factors when making decisions about grape selection, fermentation methods, and additives to minimize the presence of potentially harmful biogenic amines in their wines. In recent years, there has been growing interest in developing methods to analyze and control the levels of biogenic amines in wine. Techniques such as high-performance liquid chromatography (HPLC) and enzyme-linked immunosorbent assays (ELISA) have been used to quantify the concentrations of these compounds in wine samples. By understanding the factors that influence the production of biogenic amines and implementing strategies to monitor and control their levels, winemakers can ensure the safety and quality of their products. Research in this area can lead to the development of new tools and technologies to mitigate the risks associated with biogenic amines in wine and other food products.

Sources and Formation of Biogenic Amines

Biogenic amines are a group of nitrogen-containing compounds that can be found in various foods, including wine. These compounds play a crucial role in the sensory characteristics of wine, contributing to its flavor and aroma. Biogenic amines in wine are primarily derived from amino acids through the action of yeast during fermentation. Yeasts can decarboxylate amino acids, converting them into biogenic amines such as histamine, tyramine, and putrescine. The presence of these compounds in wine is influenced by factors such as grape variety, fermentation conditions, and aging process. The formation of biogenic amines in wine is a complex process that involves the metabolism of yeast and bacteria present during fermentation. The activity of specific strains of yeast and bacteria can impact the production of biogenic amines in wine. Certain strains of lactic acid bacteria have been found to increase the levels of biogenic amines in wine through decarboxylation of amino acids. The presence of nutrients such as nitrogen in grape must can promote the growth of bacteria that are capable of producing biogenic amines. The sources and formation of biogenic amines in wine are multifaceted and influenced by a variety of factors. Understanding the pathways by which these compounds are produced in wine is important not only for ensuring quality and safety but also for optimizing the sensory characteristics of the final product. Further research into the mechanisms of biogenic amine formation in wine is crucial for developing strategies to control and mitigate their presence, ultimately enhancing the overall quality of wine.

Health Implications of Biogenic Amines in Wine

Biogenic amines are naturally occurring compounds in wine that can have implications for human health. These compounds, such as histamine, tyramine, and putrescine, are formed during the fermentation process through the decarboxylation of amino acids by microbial enzymes. While biogenic amines are present in small concentrations in most wines, certain winemaking practices or microbial activities can lead to elevated levels, which can pose health risks to sensitive individuals. Histamine, for example, is associated with allergic reactions and can trigger symptoms such as headaches, rashes, and respiratory issues in susceptible individuals. The health implications of biogenic amines in wine are a significant concern for both consumers and winemakers. Regulations on the maximum allowable levels of these compounds in wine have been established in many countries to protect public health. Winemakers are also implementing strategies to minimize the formation of biogenic amines during winemaking, such as using selected yeast strains that do not produce these compounds or employing proper sanitation practices to control microbial populations. By understanding the factors that contribute to the formation of biogenic amines in wine and taking proactive measures to mitigate their presence, the wine industry can ensure the safety and well-being of consumers. While biogenic amines are a natural part of wine, their presence at elevated levels can have health implications for sensitive individuals. It is essential for winemakers to be aware of the potential risks associated with these compounds and to take proactive measures to prevent their formation in wine. By adhering to regulations and implementing best practices during winemaking, the industry can continue to produce high-quality wines that are safe for consumers to enjoy. Further research into the formation and mitigation of biogenic amines in wine is crucial to ensure the continued safety and health of wine drinkers around the world.

Strategies to Minimize Biogenic Amine Content in Wine

Biogenic amines are naturally occurring compounds in wine that can potentially have adverse effects on health, causing symptoms like headaches and allergic reactions in some individuals. To minimize the presence of biogenic amines in wine, winemakers can implement various strategies during the winemaking process. One approach is to carefully select and monitor the use of yeast strains, as certain strains have been found to produce higher levels of biogenic amines. By choosing yeast strains that are low in biogenic amine production or by closely regulating fermentation conditions, winemakers can reduce the overall content of these compounds in the final product. Another method to mitigate biogenic amine levels in wine is through effective sanitation practices in the winery. Contaminated equipment or inadequate cleaning procedures can introduce bacteria that contribute to biogenic amine formation during fermentation. By ensuring a stringent hygiene regime and thoroughly sanitizing all winery equipment, winemakers can minimize the risk of bacterial contamination and subsequently decrease the presence of biogenic amines in the wine. Regular monitoring of the winemaking environment for potential sources of biogenic amines can help identify and address any issues before they impact the final product. In addition to yeast selection and sanitation practices, winemakers can also employ techniques such as micro-oxygenation and the use of additives like potassium sorbate to reduce biogenic amine content in wine. Micro-oxygenation during aging can help promote the degradation of biogenic amines, while potassium sorbate can inhibit the growth of bacteria that produce these compounds. By combining multiple strategies and maintaining a proactive approach to monitoring biogenic amine levels throughout the winemaking process, winemakers can effectively minimize the presence of these potentially harmful compounds in their wines, ensuring a safer and more enjoyable drinking experience for consumers.

XVII. WINE AND HEALTH

One aspect that has garnered significant attention in recent years is the relationship between wine consumption and health. Numerous studies have explored the potential benefits associated with moderate wine intake, particularly red wine, due to its high concentration of antioxidants. Resveratrol, a polyphenol found in grape skins, has been linked to various health benefits, including reducing inflammation and protecting against certain diseases. Other compounds in wine, such as catechins and quercetin, also exhibit antioxidant properties that can help combat oxidative stress in the body, potentially lowering the risk of cardiovascular diseases and certain types of cancer. Interestingly, the French Paradox, which refers to the relatively low incidence of heart disease in France despite a diet rich in saturated fats, has often been attributed to moderate red wine consumption. The presence of polyphenols and other bioactive compounds in wine is thought to play a role in this phenomenon by supporting cardiovascular health and lowering cholesterol levels. While the exact mechanisms behind the health benefits of wine are still being researched, it is clear that the complex interplay of chemical compounds in wine can have profound effects on human physiology and wellness. It is essential to note that while moderate wine consumption may offer certain health benefits, excessive alcohol intake can have detrimental effects on ones health. Alcohol abuse can lead to liver damage, addiction, and an increased risk of various health conditions. It is crucial for individuals to consume wine and other alcoholic beverages in moderation to reap the potential health advantages while minimizing the associated risks. By understanding the chemistry of wine and its impact on health, individuals can make informed decisions about their consumption habits and prioritize their overall well-being.

Antioxidant Properties of Wine Compounds

The antioxidant properties of wine compounds play a crucial role in safeguarding the human body against oxidative stress. Polyphenols, such as flavonoids and resveratrol, are well-known antioxidants found in wine that are thought to have various health benefits. These compounds help to neutralize free radicals, which are unstable molecules that can cause cellular damage and lead to diseases such as cancer and cardiovascular disorders. Studies have suggested that moderate consumption of wine, particularly red wine, can contribute to improved antioxidant status in the body and reduce the risk of inflammatory conditions. The antioxidant capacity of wine compounds is influenced by factors such as grape variety, winemaking practices, and aging processes. Red wines typically have higher antioxidant content than white wines due to the longer maceration process, which allows for greater extraction of polyphenols from grape skins and seeds. The oak aging of wines can enhance their antioxidant properties by imparting phenolic compounds from the wood. Understanding the impact of these variables on antioxidant levels in wine is essential for maximizing the health benefits associated with its consumption. The antioxidant properties of wine compounds not only contribute to the sensory experience of drinking wine but also offer potential health advantages. By scavenging free radicals and reducing oxidative stress, these compounds may help to protect against chronic diseases and promote overall well-being. Further research into the specific mechanisms through which wine antioxidants exert their effects will continue to shed light on the potential therapeutic applications of these compounds. As the field of wine chemistry evolves, exploring the antioxidant properties of wine compounds will remain a critical area of study with implications for both the wine industry and public health.

Health Benefits and Risks of Moderate Wine Consumption

In recent years, the health benefits of moderate wine consumption have garnered significant interest from researchers and the general public alike. Studies have shown that moderate consumption of wine, particularly red wine, may be linked to a reduced risk of cardiovascular diseases. This beneficial effect is thought to be attributable to the presence of polyphenols, such as resveratrol, which possess antioxidant properties that can help protect against the development of heart disease. The anti-inflammatory properties of certain compounds in wine may also contribute to these cardiovascular benefits. On the other hand, it is important to consider the potential risks associated with wine consumption, even in moderation. One of the primary concerns is the link between alcohol consumption and an increased risk of several types of cancer, including breast, liver, and esophageal cancer. While the mechanisms underlying this association are not fully understood, it is believed that the metabolism of alcohol in the body produces harmful byproducts that can damage cells and DNA, leading to the development of cancer. Individuals should be cautious and mindful of their alcohol intake to minimize these potential risks. While moderate wine consumption may offer certain health benefits, it is crucial to weigh these advantages against the potential risks. Individuals should be informed about the possible impacts on their health and make informed decisions based on their own personal health circumstances. It is essential to strike a balance and practice moderation when it comes to alcohol consumption, taking into account both the positive and negative aspects associated with wine consumption. Maintaining a healthy lifestyle that includes a balanced diet, regular exercise, and responsible alcohol consumption is key to promoting overall well-being and reducing the risk of chronic diseases.

Research on Wine's Potential Role in Disease Prevention

Research on wines potential role in disease prevention has garnered significant attention in recent years. Numerous studies have explored the bioactive compounds present in wine, such as resveratrol and polyphenols, which have been linked to various health benefits. Resveratrol, in particular, has been shown to possess antioxidant and anti-inflammatory properties that may help reduce the risk of cardiovascular diseases and certain types of cancer. Polyphenols, on the other hand, are known for their ability to protect cells from damage and lower inflammation levels in the body, potentially contributing to the prevention of chronic diseases. Research suggests that moderate wine consumption may play a role in promoting heart health. The presence of flavonoids in wine, which are compounds known for their cardiovascular benefits, has been associated with a reduced risk of coronary heart disease. The vasodilatory effects of wine can help improve blood flow and lower blood pressure, further supporting overall heart health. While the exact mechanisms behind these benefits are still being studied, the potential of wine to positively impact disease prevention is an intriguing area of research that warrants further exploration. Despite the promising findings, it is important to note that moderation is key when it comes to incorporating wine into a healthy lifestyle. Excessive alcohol consumption can have detrimental effects on health, outweighing any potential benefits. To fully understand the role of wine in disease prevention, more research is needed to elucidate the specific pathways through which bioactive compounds in wine exert their effects. By continuing to investigate this area, researchers may uncover new insights that could potentially lead to the development of novel therapeutic approaches for combating various diseases.

XVIII. WINE CHEMISTRY IN FOOD PAIRING

One key aspect of wine chemistry that has gained considerable attention in recent years is its role in food pairing. The interactions between the chemical components of wine and the flavors of food can significantly enhance the dining experience, creating harmonious or contrasting sensations on the palate. Understanding the chemistry behind these interactions is crucial for sommeliers and chefs alike in creating balanced and memorable pairings. The acidity of wine, for example, can either complement or cut through the richness of certain dishes, while tannins can interact with proteins in food to either enhance or diminish flavors. The aromas and flavors of wine, which are attributed to compounds such as esters, terpenes, and thiols, play a crucial role in food pairing. Matching the aromatic profile of a wine with the flavors of a dish can amplify the sensory experience, creating a synergy that elevates both the wine and the food. The complexity of wine chemistry allows for an endless array of pairing possibilities, from classic matches like red wine with steak to more innovative combinations that explore contrasting flavors and textures. By understanding the chemical basis of these interactions, experts in the culinary world can create unforgettable dining experiences that engage all the senses. The intricacies of wine chemistry provide a rich tapestry of flavors and aromas that can be expertly paired with a variety of dishes to create a sensorial journey for the palate. By delving into the chemical makeup of wine, including its acidity, tannins, aromas, and flavors, sommeliers and chefs can unlock the full potential of food and wine pairings, creating harmonious or contrasting experiences that delight diners. As the appreciation for the science behind food pairing continues to grow, the role of wine chemistry in enhancing culinary experiences will only become more prominent, shaping the way we enjoy and appreciate the art of dining.

Principles of Wine and Food Compatibility

When it comes to the principles of wine and food compatibility, understanding the interactions between the components of wine and the flavors of food is essential. One key aspect to consider is the balance of acidity in both the wine and the dish being served. High-acid wines can complement fatty or savory foods by cutting through the richness, while low-acid wines may pair better with lighter, delicate dishes. Tannins, another crucial element in wine, can interact differently with various foods – for example, tannic red wines can be enhanced by the presence of fats or proteins in meat, creating a harmonious pairing that elevates both the wine and the food. The sweetness level of the wine should be taken into account when pairing it with food. Sweet wines can be excellent choices to balance spicy or salty dishes, as the sweetness can offset these intense flavors. Conversely, dry wines can complement dishes with subtle flavors that would be overwhelmed by the sweetness of a dessert wine. Aromatics in wine, such as floral or fruity notes, can also play a significant role in matching wine with food. A wine with tropical fruit aromas could be a fantastic choice to pair with a fruity dessert like a mango sorbet, enhancing the overall sensory experience of the meal. The principles of wine and food compatibility are rooted in understanding how the key elements in wine – acidity, tannins, sweetness, and aromatics – can interact with the flavors and textures of different dishes. By considering these components and their potential reactions, one can create harmonious pairings that elevate the dining experience. Experimenting with various combinations and exploring the nuances of wine and food interactions can lead to discovering new and delightful matches that enhance both the wine and the meal. Mastering the art of pairing wine and food is a dynamic and rewarding process that adds depth and complexity to the culinary experience.

Chemical Interactions Between Wine and Food

The chemical interactions between wine and food play a crucial role in enhancing the overall dining experience. When wine is paired with specific dishes, certain chemical reactions can occur that can either complement or contrast the flavors in a harmonious way. The acidity in wine can help cut through the richness of fatty foods, such as a creamy pasta dish, balancing the flavors on the palate. The tannins present in red wine can help soften the proteins in red meats, making for a more enjoyable dining experience. The complexity of wine aromas and flavors can interact with the different components of a dish, amplifying certain taste sensations. The fruity notes in a glass of Chardonnay can enhance the sweetness of a tropical fruit salad, creating a more vibrant and well-rounded flavor profile. Understanding the chemical makeup of both the wine and the food can help guide the pairing process, ensuring that each element complements the other in a balanced and cohesive manner. The chemical interactions between wine and food are a fascinating aspect of culinary science that can elevate the dining experience to new heights. By considering the acidity, tannins, aromas, and flavors of both the wine and the dish, it is possible to create pairings that enhance the overall tasting experience. A well-chosen wine can bring out the best in a dish, bringing a symphony of flavors to the palate and leaving a lasting impression on the diner. As researchers continue to explore the intricacies of these chemical interactions, the art of wine and food pairing will only continue to evolve and captivate food enthusiasts around the world.

Enhancing Flavor Harmony Through Wine Pairing

As wine enthusiasts continue to explore the complex world of flavors and aromas, the art of wine pairing has become an essential aspect of enhancing the overall sensory experience. By selecting the right wine to complement a dish, individuals can elevate the flavors of both the food and the wine, creating a harmonious balance on the palate. This synergy is achieved by considering the acidity, sweetness, tannins, and fruitiness of the wine in relation to the flavors and textures of the food. In the realm of wine pairing, the interplay between the chemical components of wine and the ingredients of a dish is crucial in achieving a harmonious balance of flavors. A rich and fatty dish may benefit from a high-acid wine to cut through the richness, while a spicy cuisine can be enhanced by a slightly sweet wine to counterbalance the heat. Understanding the chemical properties of both the wine and the food allows for a more informed selection of pairings, resulting in a heightened dining experience. The aromatic compounds present in wine play a significant role in the pairing process, as they can either amplify or complement the flavors of a dish. A wine with floral or fruity aromas may enhance the sweetness of a dessert, while a wine with earthy or herbaceous notes can complement savory dishes. By considering the bouquet of a wine alongside its other chemical components, individuals can create a synergistic pairing that enhances the overall flavor profile of both the food and the wine.

XIX. ANALYTICAL CHEMISTRY TECHNIQUES IN WINE AUTHENTICATION

An essential aspect of the wine industry is the authentication of wine products to ensure their quality and origin. Analytical chemistry techniques play a crucial role in this process by providing reliable methods for identifying and characterizing the chemical composition of wine. One common technique used in wine authentication is gas chromatography, which separates and quantifies volatile compounds in wine samples. This method can help detect adulteration or contamination in wine products, as well as provide insights into the aroma profile of different wines. Liquid chromatography coupled with mass spectrometry is another powerful tool for analyzing the non-volatile compounds in wine, such as phenolic compounds and sugar derivatives. By utilizing these analytical techniques, wine producers and regulators can verify the authenticity of wine products and protect consumers from fraudulent practices. Nuclear magnetic resonance (NMR) spectroscopy is another valuable analytical technique used in wine authentication. NMR can provide detailed information about the molecular structure of compounds present in wine samples, allowing for the identification of specific markers that are characteristic of certain grape varieties or winemaking regions. This technique is particularly useful for detecting counterfeit wines that may claim to be from prestigious vineyards or regions. By comparing the NMR spectra of authentic wines with those of suspicious samples, analysts can quickly determine whether a wine is genuine or counterfeit. The combination of gas chromatography, liquid chromatography, and NMR spectroscopy offers a comprehensive approach to wine authentication that relies on the precise analysis of chemical components in wine. Analytical chemistry techniques are indispensable tools for ensuring the authenticity of wine products in the market. By employing gas chromatography, liquid chromatography, and NMR spectroscopy, wine producers, regulators, and consumers can confidently verify the quality and origin of wine samples. These techniques enable the detection of adulteration, contamination, and counterfeit wines, thereby safeguarding the integrity of the wine industry. Moving forward, continued advancements in analytical chemistry technologies will further enhance the capabilities of wine authentication, providing more accurate and efficient methods for verifying the chemical composition of wine products.

DNA Profiling for Grape Variety Identification

The application of DNA profiling in grape variety identification has revolutionized the wine industry by providing a precise and reliable method to authenticate grape varieties. By analyzing specific genetic markers in the grapevines DNA, researchers can accurately determine the unique genetic profile of different grape varieties. This technology has enabled winemakers to verify the authenticity of their grape sources, ensuring the integrity of their products and maintaining the distinct characteristics associated with specific grape varieties. DNA profiling allows for the identification of rare or previously unknown grape varieties, contributing to the preservation and diversification of grapevine genetic resources. In addition to authenticity and diversity, DNA profiling for grape variety identification plays a crucial role in quality control and assurance in the wine production process. By

102

confirming the grape varieties used in a particular wine blend, producers can ensure consistency in flavor profiles and characteristics, enhancing consumer trust and brand reputation. DNA profiling can help detect any accidental or intentional adulteration of grape varieties, safeguarding the quality and purity of the final wine product. This level of traceability and transparency provided by DNA profiling reinforces the credibility of the wine industry and fosters greater consumer confidence in the products they purchase. The integration of DNA profiling for grape variety identification aligns with the growing trend towards sustainability and environmental responsibility in the wine industry. By accurately tracing the origins of grape varieties, producers can make informed decisions regarding vineyard management practices, optimizing resource allocation and minimizing environmental impact. This precision in grape variety identification also enables winemakers to adapt to changing climate conditions and select grape varieties that are more resilient or better suited to specific terroirs, promoting long-term sustainability and resilience in the face of climate change. DNA profiling for grape variety identification not only enhances the quality and authenticity of wines but also contributes to a more sustainable and environmentally conscious approach to winemaking.

Isotope Analysis for Geographical Origin Verification

Isotope analysis has emerged as a powerful tool for verifying the geographical origin of wines, providing valuable information for both producers and consumers. By examining the stable isotopes of elements such as carbon, hydrogen, oxygen, and sulfur in wine samples, researchers can distinguish wines from different regions based on their unique isotopic signatures. This technique takes advantage of the fact that factors like soil composition, climate, and water sources in a specific region can influence the isotopic composition of grapes, which is ultimately reflected in the wine produced from them. The process of isotope analysis involves determining the ratio of isotopes in a sample and comparing it to reference databases that contain isotopic profiles of wines from known regions. This allows researchers to pinpoint the geographical origin of a wine with a high degree of accuracy, making it a valuable tool for combating fraud and ensuring product authenticity. Isotope analysis can also provide insights into the production practices used by winemakers, as variations in isotopic composition can be linked to factors like irrigation methods, fertilization techniques, and grape maturation processes. Isotope analysis offers a non-invasive and highly precise method for verifying the geographical origin of wines, shedding light on the complex interactions between terroir, viticulture practices, and wine chemistry. As the wine industry continues to evolve and face new challenges, isotope analysis will play an increasingly important role in ensuring transparency, quality, and authenticity in wine production. By harnessing the power of isotopic signatures, stakeholders in the wine supply chain can make more informed decisions and build trust with consumers who are increasingly interested in the provenance of the products they purchase.

Spectroscopic Methods for Detecting Wine Fraud
Spectroscopic methods have emerged as powerful tools in the fight against wine fraud, offering a non-invasive way to analyze

the chemical composition of wine. By utilizing techniques such as UV-Vis spectroscopy, infrared spectroscopy, and nuclear magnetic resonance (NMR) spectroscopy, researchers can detect adulterants, assess wine authenticity, and determine key parameters like alcohol content and sugar levels. These methods rely on the interaction of wine molecules with electromagnetic radiation, providing detailed information about the chemical structure and composition of the sample. One of the key advantages of spectroscopic methods is their ability to provide rapid and accurate results, making them ideal for screening large numbers of samples in a short period of time. UV-Vis spectroscopy, for example, can detect the presence of colorants or additives that may indicate wine fraud, while NMR spectroscopy can reveal the isotopic composition of water and ethanol, providing insights into the wines origin and production process. These techniques are not only sensitive but also highly specific, allowing for the identification of subtle differences between authentic and counterfeit wines. In addition to their analytical capabilities, spectroscopic methods also offer a cost-effective solution for detecting wine fraud, as they require minimal sample preparation and instrumentation. This makes them accessible to a wide range of stakeholders in the wine industry, from producers and distributors to regulatory bodies and consumers. By leveraging the power of spectroscopy, the wine industry can maintain the integrity of its products, protect against counterfeiters, and ensure that consumers receive genuine, high-quality wines. As technology continues to advance, spectroscopic methods will play an increasingly pivotal role in detecting wine fraud and upholding the authenticity of this beloved beverage.

XX. ETHICAL AND SUSTAINABILITY ISSUES IN WINE PRODUCTION

One of the key issues facing the wine industry today is the ethical considerations surrounding wine production. Many consumers are increasingly concerned about the environmental impact of wine production, particularly in terms of water usage, pesticides, and carbon emissions. Sustainable practices, such as organic and biodynamic farming, are gaining popularity as consumers seek out wines that are produced with minimal harm to the environment. Wineries that prioritize ethical and sustainable practices are not only appealing to environmentally conscious consumers but also help to preserve the natural resources that are essential for the production of high-quality wines. Ethical considerations extend beyond environmental impact to include issues of social responsibility and fair labor practices. The wine industry relies heavily on labor-intensive processes, from vineyard work to winemaking, and ensuring fair wages and working conditions for workers is crucial. In recent years, there have been growing concerns about labor exploitation in some winemaking regions, prompting calls for transparency and accountability in the industry. Wineries that prioritize fair labor practices not only uphold ethical standards but also contribute to the well-being of the communities in which they operate. In addition to environmental and social concerns, ethical issues in wine production also encompass aspects such as labeling transparency, health implications, and the promotion of responsible drinking. Consumers are increasingly interested in knowing more about the ingredients used in winemaking and the production methods employed by wineries. Clear and accurate labeling that discloses information about additives, allergens, and other relevant details is essential for empowering consumers to make informed choices. Promoting responsible drinking practices and raising awareness about the health risks associated with excessive alcohol consumption are important steps toward fostering a culture of moderation and well-being within the wine industry.

Organic and Biodynamic Winemaking Practices

Organic and biodynamic winemaking practices have gained popularity in recent years due to a growing interest in sustainability and natural methods of cultivation. Organic winemaking involves avoiding the use of synthetic pesticides and fertilizers in the vineyard, relying instead on organic alternatives to maintain the health of the vines. This approach not only reduces the environmental impact of winemaking but also promotes the biodiversity of the vineyard ecosystem. Organic practices often lead to healthier soils, which can result in better grape quality and flavor expression in the wine. Biodynamic winemaking takes organic practices a step further by following principles outlined by Rudolf Steiner in the early 20th century. This holistic approach views the vineyard as a self-sustaining ecosystem and emphasizes the use of natural preparations and lunar cycles to guide vineyard activities. Proponents of biodynamics believe that these practices can enhance the vitality of the vineyard, resulting in wines that truly reflect the terroir. While the scientific basis for biodynamic practices may be debated, many winemakers have reported positive results in terms of grape quality and overall vineyard health. Both organic and biodynamic winemaking practices offer unique perspectives on sustainable agriculture and have sparked important conversations within the wine industry. By focusing on natural methods of cultivation and respecting the ecological balance of the vineyard, these approaches underscore the interconnectedness of the environment and the final product. While challenges exist in implementing these practices on a larger scale, the continued interest and support for organic and biodynamic wines suggest a growing recognition of the importance of sustainable winemaking in the modern world.

Fair Trade and Social Responsibility in the Wine Industry

The wine industry has seen a growing focus on fair trade practices and social responsibility in recent years. Wineries are increasingly adopting ethical sourcing practices, ensuring fair wages and working conditions for their employees, as well as supporting local communities. By prioritizing fair trade, wineries are not only promoting the well-being of their workers but also contributing to sustainable development in the regions where they operate. This shift towards social responsibility reflects a broader awareness within the industry of the need to minimize its environmental and social footprint. Consumers are becoming more conscientious about the products they purchase, seeking out wines that align with their values of sustainability and social equity. Wineries that prioritize fair trade practices are able to distinguish themselves in a crowded market, appealing to consumers who prioritize ethical sourcing and production methods. This demand for ethically produced wines is driving a shift in the industry towards greater transparency and accountability, as wineries strive to meet the expectations of a socially conscious consumer base. Fair trade and social responsibility have become integral aspects of the wine industry, shaping the way wineries operate and market their products. By emphasizing ethical sourcing and production practices, wineries can not only improve the lives of their workers and communities but also attract a growing segment of consumers who prioritize sustainability and social equity. As the industry continues to evolve, it is likely that fair trade practices will become even more prominent, setting a new standard for responsible and environmentally conscious wine production.

Environmental Impact of Wine Production and Packaging

One significant aspect of wine production that must be considered is its environmental impact. From vineyard management to packaging, wine production can have a notable effect on the environment. The farming practices employed in grape cultivation can lead to soil erosion, pesticide runoff, and water contamination. The energy-intensive process of vinification, including fermentation, filtration, and bottling, contributes to greenhouse gas emissions. The packaging of wine, particularly the use of glass bottles, requires significant resources for manufacturing and transportation, adding to the carbon footprint of the industry. To mitigate the environmental impact of wine production, some vineyards are adopting sustainable practices such as organic or biodynamic farming. These methods reduce the use of synthetic pesticides and fertilizers, promote biodiversity, and conserve water resources. Wineries are also implementing energy-efficient technologies, such as solar panels and LED lighting, to reduce their carbon footprint. Some producers are exploring alternative packaging options, such as lightweight glass bottles, bag-in-box, or recyclable cans, to decrease the environmental impact of packaging. By embracing these sustainable practices, the wine industry can reduce its environmental footprint and contribute to a more sustainable future. The environmental impact of wine production and packaging is a complex issue that requires careful consideration and action. While the industry faces challenges in reducing its carbon footprint and environmental degradation, there are promising solutions being adopted by forward-thinking vineyards and wineries. By embracing sustainable practices in grape cultivation, vinification, and packaging, the wine industry can mitigate its impact on the environment and move towards a more sustainable future. Through innovation and awareness, the chemistry of wine can be balanced with environmental stewardship for the benefit of both consumers and the planet.

XXI. WINE LABELING REGULATIONS

An integral aspect of the wine industry is the regulation of wine labeling, which aims to provide consumers with accurate and transparent information about the product they are purchasing. These regulations govern what information must be included on a wine label, such as the grape variety, origin, alcohol content, and allergen warnings. By ensuring that these details are clearly displayed, consumers can make informed choices based on their preferences and dietary restrictions. Labeling regulations help to prevent misleading claims or deceptive marketing practices, thus protecting the integrity of the wine market. In addition to providing consumers with essential information, wine labeling regulations also serve to uphold quality standards within the industry. By requiring certain labeling criteria to be met, such as appellation of origin or vintage year, these regulations help to distinguish premium wines from generic ones. This not only benefits consumers who are seeking high-quality products but also incentivizes producers to maintain strict adherence to production standards in order to market their wines effectively. These regulations contribute to the overall reputation and credibility of the wine industry as a whole. Wine labeling regulations play a crucial role in promoting fair trade practices and ensuring compliance with international standards. Through harmonizing labeling requirements across different markets, these regulations help facilitate the global trade of wine products. By establishing a level playing field for producers and importers, regulations prevent unfair competition and protect consumers from substandard products. By promoting transparency and consistency in labeling practices, these regulations contribute to building trust and confidence in the wine industry among stakeholders worldwide. Wine labeling regulations are essential for maintaining the integrity and sustainability of the global wine market.

Legal Requirements for Wine Label Information

When it comes to wine labels, there are specific legal requirements in place to ensure that consumers are provided with accurate and important information about the product they are purchasing. One crucial aspect is the inclusion of the wines alcohol by volume (ABV) content on the label. This information helps consumers make informed decisions about their alcohol consumption and ensures that they are aware of the strength of the wine they are purchasing. The label must disclose the producer or bottler of the wine, allowing consumers to trace the origin of the product and verify its authenticity. Another key requirement for wine labels is the listing of any allergens present in the wine. This is particularly important for individuals with allergies or dietary restrictions, as it allows them to avoid potential health risks associated with consuming certain ingredients. Common allergens that must be identified on wine labels include sulfites, which are often used as preservatives in winemaking. By clearly indicating the presence of allergens, wine labels promote transparency and consumer safety, ensuring that individuals can make informed choices about the products they consume. Wine labels must also include information on the grape variety or varieties used in the production of the wine. This helps consumers understand the flavor profile and characteristics of the wine they are purchasing, as different grape varieties can impart unique flavors and aromas. By providing details on the grape varietals used, wine labels enhance the consumers overall experience by offering insights into the wines origins and potential taste profile. Legal requirements for wine label information play a crucial role in promoting transparency, consumer safety, and informed decision-making in the wine marketplace.

Appellations and Geographical Indications

In the realm of wine production, appellations and geographical indications play a crucial role in defining the origin and quality of a wine. Appellations are designated regions, typically with specific regulations regarding grape varieties, yields, and winemaking practices, that help consumers identify the provenance of a wine. Geographical indications, on the other hand, are broader classifications that protect the names of regions and prevent misuse by producers from outside the area. Together, these systems serve to preserve the authenticity and integrity of wines from specific terroirs, allowing for transparency and trust between producers and consumers. Appellations and geographical indications contribute to the overall diversity and richness of the wine industry by highlighting the unique characteristics of different regions. The concept of terroir, which encompasses the soil, climate, and topography of a particular vineyard, is a key factor in determining the quality and style of a wine. By emphasizing the distinctiveness of each terroir through appellations and geographical indications, winemakers can showcase the diverse range of flavors and aromas that arise from different environments. This not only adds value to individual wines but also promotes a deeper appreciation for the environmental factors that influence wine production. The implementation of appellations and geographical indications enhances the transparency and authenticity of the wine market, allowing consumers to make more informed choices based on the origin and quality of a wine. By highlighting the unique characteristics of different regions, these systems contribute to the overall diversity and richness of the wine industry, fostering a sense of place and identity in each bottle. Moving forward, continued support and recognition of appellations and geographical indications will be crucial in promoting sustainable practices and preserving the heritage of winemaking traditions around the world.

Label Claims and Marketing Strategies

Label claims and marketing strategies play a crucial role in the wine industry, as they have the power to influence consumer perception and purchasing decisions. Wineries often use label claims to highlight specific characteristics of their wines, such as "organic," "sustainable," or "biodynamic." These claims not only appeal to consumers who prioritize ethical and environmental factors but also differentiate the wine from competitors. Marketing strategies shape how wines are positioned in the market, whether through luxury branding, storytelling, or influencer partnerships. By creating a strong brand image and narrative, wineries can build loyalty and attract a diverse range of consumers. In addition to label claims and marketing strategies, wineries must also consider regulations and legal requirements when developing their branding and communication tactics. In some regions, there are strict rules governing the use of terms like "reserve," "old vine," or "estate grown." Wineries must ensure that their label claims are compliant with these regulations to avoid fines or loss of credibility. Misleading or exaggerated marketing claims can damage a winery's reputation and erode consumer trust. It is essential for wineries to be transparent and authentic in their communication efforts to build long-lasting relationships with consumers and maintain a positive brand image. Label claims and marketing strategies are powerful tools that can shape consumer perception and drive sales in the competitive wine industry. Wineries must carefully craft their messaging to resonate with target audiences, while also ensuring compliance with regulations and fostering trust among consumers. By leveraging the right combination of label claims and marketing tactics, wineries can effectively showcase the unique qualities of their wines, establish a strong brand presence, and ultimately succeed in a crowded marketplace.

XXII. WINE AND CULINARY ARTS

The relationship between wine and culinary arts is a dynamic and influential one that has enriched gastronomy for centuries. Chefs and sommeliers often collaborate to create pairings that enhance the flavors of both the dish and the wine, showcasing the complexity and versatility of this beverage. Wine is not just a drink; it is an ingredient that can elevate a dish to extraordinary heights, adding depth, richness, and complexity to the overall dining experience. While some traditional pairings have stood the test of time, culinary experimentation continues to push boundaries, leading to innovative combinations that delight the palate and challenge conventional norms. In the realm of culinary arts, understanding the chemistry of wine is essential for creating harmonious pairings that resonate with diners. The acidity, sweetness, tannins, and aroma compounds present in wine all play a role in how it interacts with food, either enhancing or detracting from the flavors of a dish. By knowing the chemical makeup of different wines, chefs can make informed decisions about which pairings will work best, striking a balance that elevates both the food and the wine. This attention to detail and precision is what sets exceptional culinary experiences apart, as it demonstrates a deep understanding and appreciation of the nuances that make wine and food come alive on the palate. As the boundaries of culinary arts continue to expand and evolve, so too do the possibilities for wine pairings and flavor combinations. With the advent of molecular gastronomy and avant-garde culinary techniques, chefs are exploring new ways to incorporate wine into their dishes, whether through infusions, reductions, or innovative presentations. This experimental approach to food and wine pairing invites diners on a sensory journey, challenging preconceived notions and expanding the boundaries of taste and sensation. By embracing the intersection of wine and culinary arts, chefs are able to push the boundaries of creativity, constantly innovating and redefining the gastronomic experience for a new generation of food enthusiasts.

Wine Pairing Principles in Cooking

When it comes to cooking with wine, understanding the principles of wine pairing is essential to creating harmonious and flavorful dishes. The key to successful wine pairing lies in balancing the flavors of the wine with the ingredients in the dish. One fundamental principle is to match the intensity of the wine with the flavors of the food - light wines pair well with delicate dishes, while robust wines complement hearty, savory flavors. A light and crisp Sauvignon Blanc works well with a seafood dish, enhancing the fresh and subtle flavors, whereas a bold and tannic Cabernet Sauvignon is a perfect match for a rich, meaty steak. In addition to considering the intensity of the wine and food, another important aspect of wine pairing is to balance the acidity of the wine with the acidity of the dish. High-acid wines, such as a Chardonnay or a Pinot Noir, pair well with acidic dishes like tomato-based pasta sauces or citrus-marinated seafood. The acidity in the wine can help cut through the richness of the dish and cleanse the palate, creating a well-rounded and enjoyable dining experience. Conversely, pairing a low-acid wine with a highly acidic dish can result in a clash of flavors and an unpleasant taste sensation. The principle of flavor matching can also guide wine pairing choices in cooking. By identifying the dominant flavors in a dish, one can select a wine that either complements or contrasts those flavors for a cohesive overall taste experience. A fruity and aromatic Gewürztraminer can enhance the spicy flavors in a Thai curry, while a dry and mineral-driven Chablis can provide a refreshing contrast to creamy dishes like scallops in butter sauce. Understanding these wine pairing principles can help elevate the culinary experience, allowing for a harmonious union of flavors that heighten the enjoyment of both the wine and the food.

Use of Wine in Culinary Techniques

Culinary techniques involving the use of wine have been prevalent for centuries, with the beverage serving as a versatile ingredient in cooking. Beyond being enjoyed as a standalone beverage, wines complex chemical composition adds depth and richness to dishes. One common technique is deglazing, where the fond from searing meat is loosened and flavor compounds are extracted with wine. The acidity in wine helps to balance and enhance flavors, while the sugars caramelize to provide depth and complexity to sauces. In addition to deglazing, wine is often used in marinades to tenderize meat and infuse it with flavor. The tannins in red wine, for example, can help break down proteins, resulting in a more tender and flavorful dish. The aromatics and flavors in wine can add a unique dimension to dishes, whether it be a bold red wine for a robust beef stew or a crisp white wine for a delicate seafood pasta. Understanding the chemical interactions between wine and other ingredients is key to utilizing it effectively in culinary applications. The use of wine in culinary techniques not only enhances the flavor profile of dishes but also adds a sophisticated touch to the overall dining experience. Chefs use wine in reductions to create sauces with complex layers of flavor, as the alcohol content helps to carry and blend the different components together. The diversity of wines available allows for a range of creative possibilities in cooking, from using a dry white wine to brighten up a creamy risotto to incorporating a sweet dessert wine in a poaching liquid for fruit. Wines ability to elevate dishes through its chemical components showcases its valuable role in the culinary world.

Incorporating Wine in Food Recipes

When it comes to incorporating wine in food recipes, chefs have found it to be a versatile and flavorful ingredient that can elevate the taste of dishes. Whether used in sauces, marinades, or desserts, wine adds depth and complexity to a recipe that is hard to achieve with other ingredients. A splash of red wine can enhance the richness of a beef stew, while white wine can bring a bright acidity to a seafood pasta. The alcohol in wine also helps to tenderize meat and infuse it with flavor, making it a valuable addition to many meat-based dishes. In addition to flavor enhancement, the chemical composition of wine can also play a role in how it interacts with food. The acidity in wine can help to cut through the richness of fatty dishes, while tannins can add a textural element that complements meaty flavors. Aromas in wine, such as those from oak aging or fermentation, can also add layers of complexity to a recipe, creating a more nuanced dining experience. By understanding the chemical makeup of wine, chefs can better pair it with different ingredients to create harmonious and balanced dishes. Using wine in cooking is not just about enhancing flavor, but also about show-casing the unique characteristics of different wine varietals. A Pinot Noir may bring earthy notes to a mushroom risotto, while a Riesling can add a touch of sweetness to a fruit compote. By experimenting with different types of wine in recipes, chefs can explore the vast array of flavors and aromas that wine has to offer, creating a dining experience that is both delicious and educational. Incorporating wine in food recipes is a creative way to explore the chemistry of taste and elevate the dining experience for both chefs and diners alike.

XXIII. WINE TOURISM AND EDUCATION

In recent years, wine tourism has become increasingly popular as a way for individuals to deepen their understanding and appreciation of wine. Educational experiences, such as vineyard tours, wine tastings, and even wine pairing classes, are becoming common offerings at wineries around the world. These experiences provide visitors with the opportunity to learn about the various aspects of wine production, from grape cultivation to fermentation, and how these processes influence the final product in the glass. Participating in wine education activities can significantly enhance a consumers enjoyment of wine by helping them develop a more discerning palate and a greater understanding of the complexities involved in winemaking. By learning about the different chemical components of wine, such as acids, sugars, and tannins, individuals can better appreciate the nuances of flavor and aroma present in each glass. Gaining insight into the fermentation process and the role of yeast and bacteria can provide a deeper appreciation for the science behind winemaking. Wine tourism and education can foster a deeper connection between consumers and the wine they enjoy, leading to a more enriched wine-drinking experience. By expanding their knowledge of wine production and the chemical compounds that contribute to its distinct characteristics, individuals can elevate their tasting skills and make more informed choices when selecting wines. As the demand for wine tourism continues to grow, it is evident that the marriage of education and experiential learning in the world of wine is a trend that is here to stay.

Importance of Wine Tourism for the Industry

Wine tourism plays a crucial role in the wine industry, offering a unique and immersive experience for consumers who seek to deepen their knowledge and appreciation of wine. By welcoming visitors to vineyards, wineries, and wine regions, the industry can foster a deeper connection between consumers and the product, enhancing brand loyalty and encouraging future wine purchases. Through guided tours, tastings, and educational workshops, wine tourism provides valuable opportunities for consumers to learn about the winemaking process, the various grape varietals, and the impact of terroir on wine production. Wine tourism contributes significantly to the economic sustainability of the industry, generating revenue for wineries, local businesses, and entire regions. By attracting visitors from around the world, wine tourism stimulates the local economy, creates jobs, and supports small-scale producers who may not have the resources to compete on a global scale. Wine tourism can help to diversify revenue streams for wineries, reducing their dependence on traditional distribution channels and allowing them to connect directly with consumers in a more personal and authentic way. Wine tourism serves as a means of promoting cultural exchange and fostering a deeper understanding of different wine-producing regions. By showcasing the unique characteristics of each terroir, winemakers can educate visitors about the diversity of wine styles and flavors available, encouraging them to explore new wines and broaden their palate. In this way, wine tourism not only benefits the industry economically but also enriches the cultural experiences of consumers, creating lasting memories and strong emotional connections to the world of wine. The importance of wine tourism for the industry lies in its ability to engage consumers, drive sales, and promote a greater appreciation for the art and science of winemaking.

Wine Tasting Events and Tours

Wine tasting events and tours offer enthusiasts the opportunity to experience the sensory delights of various wines in a social and educational setting. These events are often hosted by wineries, wine bars, or tasting rooms, where guests can sample a range of wines while learning about the production process, grape varieties, and regional differences that contribute to the nuances of each wine. Participants can engage in discussions with knowledgeable staff or winemakers, gaining insights into the chemistry behind wine and how factors like soil composition, climate, and fermentation techniques influence the final product. Wine tasting events provide a platform for individuals to explore their palate and preferences, expanding their understanding and appreciation of wine. Tasting different wines side by side allows for comparisons in terms of flavor profiles, acidity levels, tannin structures, and aromatic complexities. Through guided tastings, participants can develop their sensory skills, discerning subtle nuances and identifying the key elements that make a wine unique. This experiential learning not only enhances ones enjoyment of wine but also fosters a deeper connection to the rich cultural heritage and scientific intricacies that underpin the world of winemaking. Wine tours offer a more immersive experience, allowing visitors to witness firsthand the vineyards, cellars, and production facilities where wine is crafted. These tours provide a behind-the-scenes look at the winemaking process, from grape cultivation and harvesting to fermentation and aging. By exploring the physical spaces and equipment used in wineries, participants can gain a comprehensive understanding of the scientific principles and techniques involved in creating high-quality wines. Interacting with winemakers and viticulturists during tours allows for in-depth discussions about sustainable practices, technological innovations, and evolving trends in the wine industry, further enriching the educational aspect of wine tasting events and tours.

Educational Programs in Wine Studies

Educational programs in wine studies have gained significant popularity in recent years, as more people seek to deepen their understanding of this complex and fascinating industry. These programs offer a comprehensive curriculum that covers a wide range of topics, from viticulture and enology to wine business and marketing. Students can expect to learn about the different grape varieties, wine regions, and styles of wine produced around the world. They also delve into the science behind winemaking, including fermentation processes, aging techniques, and the role of microorganisms in shaping the final product. Students gain practical experience through vineyard visits, wine tastings, and internships at wineries, providing hands-on learning opportunities that can be invaluable in preparing for a career in the wine industry. Educational programs in wine studies often emphasize the importance of sustainability and ethical practices in winemaking. Students learn about organic and biodynamic farming methods, as well as the use of renewable energy and water conservation techniques in vineyard management. These programs also highlight the cultural and historical significance of wine, exploring its role in society, religion, and art throughout the ages. By incorporating these broader perspectives into the curriculum, students gain a more holistic understanding of the wine industry and its impact on the environment and society. This interdisciplinary approach helps prepare students for the diverse challenges and opportunities they may encounter in their future careers in the wine industry. Educational programs in wine studies play a crucial role in developing the next generation of wine professionals and enthusiasts. By providing a comprehensive and multidisciplinary education, these programs equip students with the knowledge and skills needed to succeed in a competitive and rapidly evolving industry. Whether one aims to become a winemaker, sommelier, viticulturist, or wine marketer, these programs offer a solid foundation upon which to build a successful career in the world of wine. With a growing demand for skilled professionals in the wine industry, investing in quality education through these programs can be a strategic and rewarding decision for individuals passionate about wine and eager to make a meaningful impact in the field.

XXIV. WINE MARKETING AND CONSUMER TRENDS

Consumer trends in the wine industry play a significant role in shaping the marketing strategies employed by wineries around the world. With the rise of social media and online platforms, consumers now have access to a wealth of information on different wine varietals, regions, and producers. This means that wineries must not only produce high-quality wines but also effectively communicate their unique selling points to a more knowledgeable and discerning consumer base. In response to this shift, wineries are increasingly focusing on storytelling, sustainability, and authenticity as key factors in their marketing campaigns. One notable trend in wine marketing is the emphasis on sustainability and environmental responsibility. Consumers are becoming more conscious of the environmental impact of their purchasing decisions, leading many wineries to adopt organic, biodynamic, or sustainable farming practices. This not only appeals to environmentally conscious consumers but also adds a layer of authenticity and transparency to the brand. Wineries that can demonstrate a commitment to sustainability can attract a loyal customer base willing to pay a premium for ethically produced wines. The emergence of experiential marketing in the wine industry has been a game-changer in attracting consumers and building brand loyalty. From vineyard tours and tastings to wine pairing events and educational seminars, wineries are increasingly offering immersive experiences to engage with consumers on a deeper level. By creating memorable experiences around their brand, wineries can forge a stronger emotional connection with consumers, leading to increased brand loyalty and advocacy. In an increasingly competitive marketplace, wineries that can offer unique and engaging experiences are likely to stand out and capture the attention of discerning wine enthusiasts.

Strategies for Promoting Wine Brands

One effective strategy for promoting wine brands is to utilize social media platforms to engage with consumers and build brand awareness. By sharing captivating content such as vineyard tours, behind-the-scenes footage of the winemaking process, and food pairing suggestions, wineries can create an interactive experience for their audience. Hosting virtual tastings or live Q&A sessions can further enhance consumer engagement and loyalty. Social media platforms also allow wineries to target specific demographics through tailored advertisements, ensuring that their promotions reach the right audience. Another key strategy for promoting wine brands is to participate in wine competitions and tastings. Winning prestigious awards or receiving high ratings from renowned critics can significantly boost a wineries reputation and credibility in the industry. By showcasing these accolades on their labels, websites, and marketing materials, wineries can effectively differentiate themselves from competitors and attract discerning consumers. Participating in wine festivals and events provides wineries with valuable opportunities to showcase their products to a diverse audience and connect with potential customers on a personal level. Collaborating with influencers and brand ambassadors can also be a powerful strategy for promoting wine brands. Partnering with well-known personalities in the food and beverage industry or social media influencers with a strong following can help wineries reach a larger audience and increase brand visibility. By leveraging the credibility and influence of these individuals, wineries can effectively promote their products and connect with new customers who may be swayed by the endorsement of a trusted source. Building strong relationships with influencers can lead to long-term brand advocacy and sustainable growth in a competitive market.

Consumer Preferences and Market Research

Consumer preferences play a crucial role in shaping the wine market, influencing everything from production to marketing strategies. Market research is essential for understanding these preferences, helping producers determine the types of wine consumers prefer, the price points they are willing to pay, and the packaging and labeling that appeal to them. By analyzing consumer data, producers can tailor their offerings to meet the demands of the market, ensuring that they remain competitive and relevant. This in-depth understanding of consumer preferences can lead to increased sales and brand loyalty. Market research can also provide insights into emerging consumer trends, allowing producers to anticipate shifts in demand and adjust their offerings accordingly. As more consumers seek out organic and sustainable products, wineries can use market research to identify opportunities to expand their organic wine portfolios. By staying attuned to changing consumer preferences, producers can stay ahead of the curve and maintain their position in a dynamic and competitive market. Market research thus serves as a valuable tool for navigating the ever-evolving wine industry landscape. In addition to understanding consumer preferences, market research can also help producers identify new target markets and opportunities for growth. By analyzing market trends and consumer behaviors, producers can uncover niche markets that may be underserved or untapped. This can lead to the development of new products tailored to specific consumer segments, opening up new avenues for revenue and market expansion. Through strategic market research, producers can position themselves for success in a competitive market environment, where understanding consumer preferences is key to staying relevant and profitable.

Influence of Social Media on Wine Sales

Social media has become a powerful tool for marketing and promoting products, including wine. Platforms like Instagram, Facebook, and Twitter have enabled wineries to reach a wider audience and engage with consumers in ways that were previously impossible. By sharing stunning images of vineyards, behind-the-scenes glimpses of the winemaking process, and interactive content like virtual wine tastings, wineries can pique the interest of potential customers and increase brand awareness. These platforms also allow for direct communication with consumers, providing a more personalized and engaging experience that can foster loyalty and drive sales. Social media influencers and wine bloggers have become key players in shaping consumer perceptions and influencing purchasing decisions. With large followings and a strong presence in the online wine community, influencers can significantly impact the popularity of certain wines or regions. Their recommendations, reviews, and sponsored content can sway consumer preferences and drive sales for specific brands. This phenomenon highlights the shifting landscape of wine marketing, as traditional advertising methods are being replaced by more organic and authentic endorsements from trusted voices in the industry. Social media provides valuable data and insights that wineries can use to optimize their marketing strategies and target their audience more effectively. Analytics tools allow businesses to track consumer behavior, monitor trends, and measure the success of their social media campaigns. By analyzing this data, wineries can identify which types of content resonate most with their audience, refine their messaging, and tailor promotions to better meet consumer preferences. This data-driven approach can lead to increased sales, brand loyalty, and a stronger online presence in an increasingly competitive market.

XXV. WINE INDUSTRY REGULATIONS

As the wine industry continues to evolve, regulations play a crucial role in ensuring quality, safety, and fair trade practices within the sector. These regulations encompass a wide range of factors, including labeling requirements, production standards, geographical indications, and marketing practices. In order to protect consumers from misleading information, wine labels often have to adhere to strict guidelines regarding grape varietals, vintage, and alcohol content. Regulations regarding production standards ensure that winemakers follow specific processes to maintain the integrity of the final product. Geographical indications (GIs) are another key aspect of wine industry regulations, as they establish the link between a specific region and the quality, reputation, and characteristics of the wine produced there. GIs are designed to protect traditional winemaking practices and prevent producers outside the region from unfairly benefiting from the reputation of a specific terroir. By enforcing these regulations, governing bodies help maintain the authenticity and uniqueness of wines from different regions, preserving their cultural heritage and promoting sustainable viticulture practices. In addition to traditional regulations, the wine industry is also facing new challenges related to climate change, technological advancements, and shifting consumer preferences. As a result, regulatory bodies are constantly adapting to these changes by updating standards, incorporating new technologies, and addressing emerging issues such as sustainability and organic practices. These evolving regulations not only shape the future of the wine industry but also reflect the dynamic nature of an industry that is deeply rooted in tradition yet constantly seeking innovation and improvement.

Government Oversight of Wine Production

Government oversight of wine production plays a crucial role in ensuring the quality, safety, and authenticity of wines available to consumers. Regulatory bodies, such as the Alcohol and Tobacco Tax and Trade Bureau (TTB) in the United States, set strict standards regarding labeling, production methods, and ingredient use in winemaking. These regulations help protect consumers from fraudulent practices and ensure that wines meet certain quality benchmarks. In addition to quality control, government oversight also extends to environmental regulations and sustainability practices in the wine industry. Many countries have regulations in place to monitor the use of pesticides, water usage, and waste management in winemaking processes. By enforcing these regulations, governments can help reduce the environmental impact of wine production and promote sustainable practices within the industry. Government oversight of wine production serves to maintain a level playing field for winemakers, protect consumer interests, and promote sustainable practices within the industry. By setting standards for labeling, production methods, and environmental regulations, regulatory bodies play a key role in shaping the future of the wine industry and ensuring its continued growth and success. The partnership between government oversight and winemakers is essential in creating a thriving and responsible wine industry that prioritizes quality, authenticity, and sustainability.

Compliance with Alcohol Laws and Regulations

One critical aspect of the wine industry is ensuring compliance with alcohol laws and regulations. The legal framework surrounding the production, distribution, and sale of wine plays a crucial role in maintaining the integrity of the industry and protecting consumers. Strict regulations dictate everything from labeling requirements to alcohol content limits, and adherence to these rules is essential for businesses operating in the wine market. Failure to comply with alcohol laws can result in hefty fines, loss of licenses, and reputational damage, making it imperative for wineries to stay abreast of regulatory changes and requirements. In order to navigate the complex landscape of alcohol laws, wineries often invest in legal counsel to ensure their operations are in full compliance with state and federal regulations. Legal experts can provide guidance on everything from licensing procedures to tax obligations, helping wineries avoid costly legal pitfalls and maintain a good standing within the industry. Compliance with alcohol laws also extends to marketing and advertising practices, with restrictions in place to prevent misleading claims and promote responsible consumption. By following these guidelines, wineries can build trust with consumers and uphold the integrity of the industry as a whole. Compliance with alcohol laws and regulations is a cornerstone of the wine industries sustainability and growth. By adhering to legal requirements, wineries can operate ethically and responsibly, contributing to a positive reputation for the industry as a whole. By upholding regulatory standards, wineries demonstrate their commitment to quality, safety, and consumer protection, fostering a culture of trust and transparency within the market. As the wine industry continues to evolve, maintaining compliance with alcohol laws will remain a critical component of success for businesses looking to thrive in a competitive and regulated environment.

Trade Agreements and Tariffs Impacting Wine Trade

Trade agreements and tariffs play a significant role in shaping the international wine trade. When countries negotiate trade agreements, they often include provisions that affect the import and export of wine. These agreements can impact tariffs, quotas, and regulations, which in turn influence the flow of wine between countries. A trade agreement between two countries may reduce or eliminate tariffs on wine imports, making it more affordable for consumers and boosting sales for wine producers in both countries. Tariffs on wine imports can directly impact the price of wine for consumers. High tariffs can make imported wines more expensive, potentially limiting consumer choice and impacting the competitiveness of foreign wine producers in the domestic market. Trade agreements can also affect the labeling and certification requirements for imported wines, which can influence consumer perceptions and purchasing decisions. Understanding the implications of trade agreements and tariffs is essential for wine producers and policymakers to navigate the global wine market successfully. Trade agreements and tariffs are crucial determinants of the dynamics of the wine trade. By shaping the conditions under which wine is imported and exported, these agreements have far-reaching implications for wine producers, consumers, and the industry as a whole. It is essential for stakeholders in the wine trade to stay informed about the evolving landscape of trade agreements and tariffs in order to adapt to changes and seize new opportunities for growth and expansion in the global marketplace. A nuanced understanding of the impact of trade agreements and tariffs is vital for a thriving and sustainable wine trade ecosystem.

XXVI. WINE AND CULTURAL HERITAGE

One of the most fascinating aspects of wine is its connection to cultural heritage. Throughout history, wine has played a significant role in various civilizations, symbolizing celebrations, rituals, and social gatherings. The cultivation of vines and the art of winemaking have been passed down through generations, preserving a rich tradition that reflects the cultural identity of different regions. In many countries, wine production is deeply intertwined with local customs, beliefs, and values, making it not just a beverage but a symbol of heritage. The cultural significance of wine is also evident in the way it is consumed and appreciated. In many societies, the act of sharing a glass of wine is a social ritual that fosters connections between individuals and strengthens community bonds. Wine also often features prominently in religious and ceremonial practices, underscoring its spiritual importance. The imagery and symbolism associated with wine in art, literature, and folklore further highlight its enduring role as a symbol of cultural heritage. The diverse range of wines produced around the world reflects the myriad of cultural influences that have shaped winemaking practices. Each wine region has its unique terroir, grape varietals, and winemaking techniques that are influenced by local traditions and customs. By exploring the cultural heritage of wine, we not only gain a deeper appreciation for this ancient beverage but also a better understanding of the diverse societies and civilizations that have contributed to its evolution over centuries. Wine serves as a bridge between the past and the present, connecting us to our heritage and cultural roots.

Historical Significance of Wine in Different Cultures

One of the most fascinating aspects of the historical significance of wine in different cultures is the integral role it has played in religious ceremonies and rituals. In ancient Egypt, wine was associated with the god of the afterlife, Osiris, and was used in funerary rituals to guide the deceased to the underworld. Similarly, in ancient Greece, wine was central to the worship of Dionysus, the god of wine, fertility, and revelry. The Greeks believed that sharing wine during ceremonies helped to bring people together and establish social bonds. Wine has been a symbol of status and wealth in many cultures throughout history. In ancient Rome, for example, the elite classes showcased their wealth by consuming expensive and rare wines at banquets and feasts. The quality and quantity of wine served were seen as indicators of social standing and sophistication. This tradition continued in medieval Europe, where wine was reserved for nobility and clergy, further cementing its association with power and privilege. In addition to its religious and social significance, wine has also been used for medicinal purposes in various cultures. Ancient civilizations such as the Chinese, Egyptians, and Greeks believed in the healing properties of wine when consumed in moderation. It was thought to aid digestion, improve circulation, and even cure certain ailments. This belief in the therapeutic benefits of wine persisted throughout the ages and influenced the development of early pharmacology. The historical importance of wine in different cultures is a testament to its enduring legacy and its deep-rooted connection to human society.

Rituals and Traditions Involving Wine

Ceremonies and traditions involving wine have a long history dating back to ancient civilizations. In many cultures, wine plays a central role in religious ceremonies, symbolizing the blood of Christ or the essence of life itself. These rituals often involve specific ways of serving and consuming wine, with attention paid to the vessels used, the manner of pouring, and the words spoken during the ceremony. These traditions can vary widely between cultures, but they all serve to elevate the act of wine drinking to a sacred or symbolic level. Wine has been a part of social rituals and celebrations for centuries, bringing people to-gether in a shared experience of enjoyment and conviviality. From toasting at weddings to sharing a glass of wine during festive gatherings, wine has a way of fostering camaraderie and enhancing the sense of community. The act of pouring wine for others or raising a glass in a toast can signify respect, friendship, and goodwill, creating a sense of connection and belonging among those partaking in the ritual. In addition to religious and social rituals, the production and consumption of wine have also given rise to a rich tapestry of traditions that vary from region to region. From the harvesting of grapes by hand in certain vine-yards to the use of specific oak barrels for aging wine, each step of the winemaking process can be steeped in tradition and rit-ual. These practices are often based on centuries-old knowledge passed down through generations, reflecting a deep respect for the land, the vines, and the craft of winemaking. Rituals and traditions involving wine serve to deepen our appreciation for this complex and nuanced beverage, connecting us to the past while shaping the future of wine culture.

Wine's Role in Celebrations and Festivals

Throughout history, wine has played a significant role in various celebrations and festivals across cultures. In many societies, wine is a symbol of abundance, prosperity, and social cohesion. During important events such as weddings, religious ceremonies, or harvest festivals, the presence of wine often enhances the sense of joy and unity among participants. The act of sharing a glass of wine can create a bond between individuals, fostering a sense of community and togetherness. In addition to its symbolic significance, the chemical composition of wine itself contributes to its role in celebrations and festivals. The complex interplay of acids, sugars, tannins, and aromatic compounds creates a plethora of flavors and aromas that can enhance the sensory experience of any gathering. Whether it is the crisp acidity of a white wine or the bold tannins of a red, the diversity of wine styles offers a range of options to suit different palates and occasions. The act of wine consumption during celebrations and festivals can be a form of sensory exploration and appreciation. By savoring the nuances of different wines, individuals can engage with their senses and develop a deeper understanding of the complexities of winemaking. This process of sensory exploration can elevate the experience of a celebration, turning a simple gathering into a memorable and enriching event. In this way, wine becomes not just a beverage, but a vehicle for cultural expression and sensory pleasure during moments of communal joy and festivity.

XXVII. WINE AND ARTISTIC EXPRESSION

Artistic expression in the realm of wine extends beyond the liquid itself to encompass the labels, packaging, and overall branding of a particular wine. Winemakers often collaborate with artists to create visually appealing labels that capture the essence and story of the wine within. These labels serve as a form of artistic expression, providing consumers with a glimpse into the winemakers vision and the unique characteristics of the wine. The design elements, color choices, and imagery on a wine label can evoke emotions and set the tone for the tasting experience, making it a crucial aspect of wine marketing and consumer appeal. The artistry of wine extends to the sensory experience it offers. Wine tasting can be seen as a form of artistic expression, as individuals use their senses to appreciate the complexity and nuances of different wines. The aromas, flavors, textures, and colors of wine all play a role in creating a multi-sensory experience that engages not only the palate but also the mind and emotions. Wine enthusiasts often use descriptive language to articulate their tasting notes, drawing parallels to art forms such as painting, music, or literature to convey the beauty and complexity of the wine they are experiencing. In essence, wine can be viewed as a medium for artistic expression, where winemakers, artists, and consumers alike are able to showcase their creativity and individuality. From the visual appeal of a well-designed label to the sensory delight of a finely crafted wine, each aspect of the wine experience offers opportunities for artistic interpretation and expression. By exploring the intersection of wine and art, we can gain a deeper appreciation for the cultural significance and creative possibilities inherent in this beloved beverage.

Wine as a Muse for Artists

Not only does wine hold a special place in our cultures and traditions, but it also has the unique ability to inspire artists across various disciplines. For centuries, painters, poets, musicians, and writers have found inspiration in the beauty of wine, the rituals associated with its consumption, and the deep emotional connections it evokes. Artists often use wine as a muse to capture moments of joy, passion, and melancholy, as well as to express the complexities of human relationships and the passage of time. In painting, wine has been a recurring motif, symbolizing abundance, celebration, and indulgence. Artists like Henri Matisse and Vincent van Gogh used wine as a subject in their still-life paintings, exploring its vibrant colors, rich textures, and cultural significance. The swirling patterns of a glass of red wine or the tranquil scene of a vineyard at sunset have inspired countless works of art, each capturing the essence of this ancient beverage. Similarly, writers and poets have written odes to wine, celebrating its intoxicating effects and its power to bring people together in friendship and love. Musicians have also drawn inspiration from wine, often composing pieces that evoke the sensory experience of tasting a fine vintage or the convivial atmosphere of a lively gathering. From classical symphonies to modern jazz compositions, the music inspired by wine reflects the complexity, elegance, and depth of this multifaceted beverage. As artists continue to explore the myriad ways in which wine influences our senses and emotions, it is clear that this cherished drink will remain a muse for creative expression for generations to come.

Wine Label Design and Artwork

Wine label design and artwork play a crucial role in captivating consumers attention and conveying the essence of the wine inside the bottle. A well-designed label can evoke emotions, tell a story, or create a sense of luxury, ultimately influencing the consumers perception and purchasing decision. The use of colors, typography, imagery, and branding elements all contribute to the overall aesthetic appeal of the label. Elegant script fonts and gold foil accents may suggest a high-end, sophisticated wine, while vibrant colors and playful illustrations may target a younger, more adventurous audience. Wine label design can provide valuable information to consumers, such as the grape varietal, vintage, region, and winery, helping them make informed choices. The design elements and visual cues on the label can also communicate the wines style and flavor profile, whether it is crisp and citrusy or bold and full-bodied. Labels can showcase certifications, awards, or sustainable practices, further influencing consumers who prioritize environmental and ethical considerations in their purchasing decisions. Thus, wine label design serves as a powerful marketing tool that enhances brand recognition and establishes a connection with consumers. In a competitive market where hundreds of wines vie for consumers attention, a well-executed label design can set a wine apart from its competitors and create a lasting impression. Brands that invest in unique, creative, and visually appealing labels are more likely to stand out on crowded shelves and resonate with discerning consumers. Innovative wine label designs that challenge traditional norms or incorporate interactive elements can create memorable experiences and foster brand loyalty. As the wine industry continues to evolve, the role of design and artwork on wine labels will remain pivotal in shaping consumers perceptions and driving sales.

Wine-Inspired Literature and Music

Exploring the interplay between wine, literature, and music provides a rich tapestry of artistic expressions that reflect the complexity and depth of the wine experience. In literature, wine has been a recurring motif symbolizing celebration, revelry, and intoxication. From the biblical references to wine in the Last Supper to the poetic verses of Rumi extolling the virtues of wine as a metaphor for spiritual enlightenment, writers have drawn inspiration from the sensual pleasures and social rituals associated with wine. This connection between wine and literature highlights the role of wine as a catalyst for creativity and reflection, shaping narratives and characters in works of fiction. Similarly, the relationship between wine and music is evident in the lyrical compositions that capture the nuances of taste, aroma, and texture found in a glass of wine. Musical pieces inspired by the sensory experience of wine often evoke feelings of nostalgia, romance, or melancholy, mirroring the emotional complexity associated with wine consumption. Whether it is the jazzy notes of a saxophone echoing the effervescence of a sparkling wine or the melancholic strains of a violin evoking the bittersweet taste of a tannic red, music has the power to transport listeners to the sensory landscape of wine. This fusion of wine and music creates a sensory symphony that engages both the palate and the ear, enhancing the overall experience of indulging in wine. In essence, wine-inspired literature and music serve as artistic reflections of the cultural and sensory dimensions of wine, capturing the essence of this complex beverage through the lens of words and melodies. By exploring the themes, symbols, and emotions associated with wine in literature and music, one can gain a deeper understanding of the profound impact that wine has had on human expression and creativity throughout history. From the romantic allure of a glass of Bordeaux in a sonnet to the soulful melodies inspired by the aromas of a Burgundy, wine continues to be a muse for artists seeking to evoke the sensual pleasures and profound mysteries of this ancient elixir.

XXVIII. WINE AND TECHNOLOGY INTEGRATION

In recent years, there has been a significant trend towards integrating technology into the winemaking process to improve efficiency, quality, and consistency. One notable application of technology in the wine industry is the use of precision viticulture techniques, which involve the deployment of sensors, drones, and satellite imaging to monitor vineyard conditions and optimize grape production. By collecting data on soil moisture levels, canopy density, and vine health, winemakers can make more informed decisions regarding irrigation, fertilization, and harvesting, ultimately leading to better grape quality and higher yields. Another key area where technology is making an impact in the wine industry is in the cellar, where modern winemaking equipment and techniques are revolutionizing the fermentation and aging processes. Temperature-controlled stainless steel tanks have replaced traditional wooden barrels in many wineries, allowing for more precise control over fermentation temperatures and better preservation of fruit aromas. The use of specialized yeasts and enzymes has become commonplace in winemaking to enhance flavor extraction, improve fermentation kinetics, and stabilize wine quality. These technological advancements have not only streamlined production processes but also enabled winemakers to experiment with new styles and flavors. Technology is reshaping the way wine is marketed, sold, and consumed, with online platforms, mobile apps, and social media playing an increasingly important role in connecting consumers with producers. Virtual wine tastings, augmented reality labels, and personalized wine recommendations are just a few examples of how technology is enhancing the wine-drinking experience and expanding the reach of wineries to a global audience. As the wine industry continues to embrace and adapt to technological innovations, it is clear that the integration of technology is not only here to stay but also poised to drive further growth and innovation in the future.

Innovations in Wine Production Technology

Advancements in wine production technology have played a significant role in shaping the modern wine industry. One notable innovation is the use of AI and machine learning in vineyard management. By analyzing data on soil composition, weather patterns, and grape quality, these technologies can help vineyard managers make more informed decisions to optimize grape production. This not only improves the quality of the grapes but also increases efficiency and sustainability in the vineyard. Another key innovation in wine production technology is the use of specialized fermentation vessels. Traditional oak barrels have been supplemented with stainless steel tanks, concrete eggs, and amphorae made of clay. These vessels offer winemakers greater control over the fermentation process, allowing for more precise adjustments to temperature, oxygen exposure, and microbial activity. This innovation has led to the development of wines with unique textures, aromas, and flavors that were previously unattainable with traditional winemaking methods. The adoption of state-of-the-art filtration and purification techniques has revolutionized the way wines are clarified and stabilized. Technologies such as cross-flow filtration, centrifugation, and reverse osmosis have allowed winemakers to remove impurities, reduce microbial contamination, and extend the shelf life of their wines without compromising flavor or aroma. These advancements have not only improved the overall quality of wines but also enabled producers to meet the demands of a global market with consistently high standards. Innovations in wine production technology have brought about transformative changes in vineyard management, fermentation processes, and wine clarification methods, paving the way for a new era of excellence in winemaking.

Digital Platforms for Wine Sales and Distribution

In the modern age, digital platforms have revolutionized the way wine is sold and distributed. These platforms provide wineries with a direct channel to reach consumers, cutting out the middlemen and traditional distribution networks. By selling online, wineries can expand their customer base beyond their local market and reach a global audience. These platforms also offer consumers a convenient way to purchase wine without having to visit a physical store, making it easier for them to discover and purchase new wines from different regions. Digital platforms for wine sales and distribution offer wineries valuable data and insights into consumer preferences and behavior. By collecting and analyzing data on customer purchases and preferences, wineries can tailor their marketing strategies and product offerings to better meet the needs and desires of their target audience. This data-driven approach not only helps wineries increase sales but also allows them to build stronger relationships with their customers by providing personalized recommendations and offers. Digital platforms streamline the logistics of wine distribution, making it easier for wineries to manage inventory, fulfillment, and shipping. By leveraging technology such as automated inventory management systems and integrated shipping solutions, wineries can reduce costs and improve efficiency in the distribution process. This not only benefits wineries by lowering operational expenses but also enhances the overall customer experience by ensuring fast and reliable delivery of their wines. Digital platforms play a crucial role in modernizing the wine industry and adapting to the changing preferences and demands of consumers in a digital age.

Virtual Reality Experiences in Wine Industry

Virtual reality (VR) experiences have begun to revolutionize the wine industry by offering interactive and immersive ways for consumers to learn about and engage with different wine products. Through VR technology, wine enthusiasts can virtually tour vineyards, explore wine cellars, and participate in virtual tastings from the comfort of their own homes. This not only adds a new level of convenience for consumers but also enhances their overall understanding and appreciation of the winemaking process. By simulating real-life experiences, VR can help consumers develop a deeper connection with the wine they are consuming. One of the key benefits of incorporating VR experiences in the wine industry is the ability to provide consumers with a more personalized and tailored approach to learning about wine. Through VR simulations, consumers can customize their virtual wine tasting experiences based on their preferences, allowing them to explore different grape varieties, regions, and winemaking techniques in a more engaging and interactive way. This level of personalization can help consumers discover new wines that align with their tastes and preferences, ultimately leading to a more enjoyable and satisfying wine-drinking experience. VR experiences in the wine industry have the potential to bridge the gap between traditional and digital marketing strategies, offering wineries new and innovative ways to reach and engage with consumers. By creating captivating and immersive VR experiences, wineries can effectively showcase their brand story, winemaking processes, and unique selling points to a wider audience. This not only helps wineries differentiate themselves in a competitive market but also fosters greater consumer loyalty and brand awareness. As technology continues to advance, VR experiences are poised to play an increasingly important role in shaping the future of the wine industry, offering endless possibilities for enhancing consumer education, engagement, and overall enjoyment of wine.

XXIX. WINE INVESTMENT AND COLLECTING

The world of wine investment and collecting is a fascinating realm that combines the appreciation of fine wines with the potential for financial gain. Many individuals are drawn to the idea of building a wine collection not just for personal enjoyment but also as a form of investment. The value of fine wines can appreciate over time, making them a sought-after alternative asset. Investing in wine requires a deep understanding of the market, including factors like producer reputation, vintage quality, and market trends. Collectors must balance their passion for wine with the need to make strategic decisions to ensure their investment yields a return in the long run. Successful wine investors often rely on the expertise of wine professionals, critics, and auction houses to guide their purchasing decisions. These experts can provide valuable insights into the potential value of a particular wine or vintage, helping investors navigate the complexities of the market. Collectors must carefully consider factors like storage conditions, provenance, and authenticity when acquiring wines for investment purposes. Proper storage is crucial to preserving the quality and value of a wine collection, as improper conditions can lead to deterioration and loss of value. By investing in high-quality storage facilities and maintaining meticulous records of their collection, wine investors can safeguard their assets and maximize their potential for appreciation. Wine investment and collecting present a unique opportunity for individuals to combine their passion for wine with the potential for financial growth. While the world of wine can be complex and unpredictable, with proper research, guidance, and attention to detail, investors can make informed decisions that will yield rewards over time. By understanding the market, leveraging expert advice, and prioritizing the proper care and storage of their collections, wine enthusiasts can enjoy the benefits of both a valuable investment portfolio and a curated selection of exceptional wines to savor and share with others.

Economics of Wine Investment

One aspect of the wine industry that has garnered attention in recent years is the economics of wine investment. Historically, wine has been perceived not only as a beverage but also as a valuable asset with the potential for financial gain. The allure of wine as an investment lies in its ability to potentially provide high returns compared to traditional investments such as stocks or bonds. Wine investors often look for rare and highly sought-after bottles that can appreciate in value over time, especially as they age and become increasingly scarce in the market. The value of wine as an investment is influenced by a variety of factors, including market trends, producer reputation, vintage quality, and global demand. Wines from renowned regions like Bordeaux or Burgundy tend to command higher prices due to their established reputation for quality and aging potential. The vintage year can significantly impact the value of a wine, with exceptional years often fetching higher prices in the market. Investors must carefully consider these factors and conduct thorough research before making decisions to purchase wine for investment purposes. Despite its potential for high returns, wine investment also poses risks and challenges that investors must navigate. Market volatility, changing consumer preferences, counterfeit wines, and storage conditions can all impact the value and authenticity of wine investments. Wine is a tangible asset that requires proper storage and maintenance to preserve its quality and value over time. As such, investors in the wine market must possess a combination of market knowledge, financial acumen, and passion for wine to navigate the complexities of wine investment successfully.

Factors Influencing Wine Collecting Trends

One of the key factors influencing wine collecting trends is the increasing popularity of wine as an investment vehicle. In recent years, wine has gained traction as an alternative investment, with collectors viewing it as a tangible asset that has the potential to increase in value over time. This trend is fueled by the limited supply of rare and sought-after wines, driving prices up in the secondary market. As a result, investors and collectors are drawn to wine not only for its sensory appeal but also for its potential financial returns. The intersection of wine appreciation and financial gain has led to a growing interest in wine collecting as a form of investment. Another factor shaping wine collecting trends is the influence of social media and digital platforms on consumer behavior. With the rise of wine influencers, bloggers, and online communities dedicated to wine culture, consumers have greater access to information about different wines, regions, and producers. This easy access to knowledge has empowered collectors to make more informed decisions about their wine purchases and collections. Social media has created a platform for collectors to share their passion for wine, connect with like-minded individuals, and showcase their collections. The visibility and social validation that come with sharing wine-related content online have contributed to the growth of wine collecting as a popular hobby among a diverse range of enthusiasts. Evolving consumer preferences and tastes play a significant role in shaping wine collecting trends. As consumers become more knowledgeable about wine and seek out unique and distinctive expressions, there is a growing interest in exploring wines from lesser-known regions, experimenting with natural and biodynamic wines, and discovering new grape varietals. This shift towards discovery and exploration in the world of wine has fueled demand for wines that offer a sense of authenticity, terroir-driven flavors, and a connection to sustainable and artisanal winemaking practices. Collectors are increasingly drawn to wines that tell a story, reflect a sense of place, and offer a glimpse into the winemakers philosophy. As a result, the market for collectible wines has expanded beyond traditional categories, reflecting a diverse and dynamic landscape driven by evolving consumer preferences.

Wine Auctions and Valuation Methods

Wine auctions play a significant role in the wine industry, providing a platform for rare and fine wines to be sold to collectors and enthusiasts around the world. One key aspect of wine auctions is the valuation methods used to determine the value of a particular bottle or lot. Auction houses typically rely on a combination of factors to assess the worth of a wine, including provenance, condition, rarity, vintage, producer, and critical acclaim. These factors are crucial in determining the estimated price range at which a wine should be auctioned, ensuring that both buyers and sellers are satisfied with the final transaction. Valuation methods in wine auctions often involve the expertise of experienced wine professionals, such as sommeliers, collectors, and wine critics, who can assess the quality and value of a specific bottle based on their knowledge and tasting experience. Market trends and demand for certain wines can heavily influence the valuation process, as wines that are currently in vogue or highly sought after may command higher prices at auction. Understanding the nuances of these valuation methods is essential for both buyers and sellers participating in wine auctions, as it can help them make informed decisions and secure the best possible outcome for their transactions. Wine auctions and their associated valuation methods are integral components of the wine market, facilitating the exchange of rare and sought-after wines among collectors and connoisseurs. By considering factors such as provenance, condition, rarity, vintage, producer, and market demand, auction houses can accurately determine the value of a wine and set appropriate price estimates for auction. The expertise of wine professionals and the evolving trends within the industry further contribute to the valuation process, highlighting the dynamic nature of the wine market and the importance of staying informed in order to succeed in wine auctions.

XXX. WINE AND GLOBALIZATION

Globalization has had a significant impact on the wine industry, leading to an increase in the production and consumption of wine on a global scale. With advancements in transportation and technology, wines from different regions are now readily available worldwide, allowing consumers to explore a diverse range of flavors and styles. This interconnectedness has also led to the exchange of winemaking techniques and knowledge between different wine-producing regions, resulting in the growth of new wine markets and the emergence of innovative wine styles. One of the key ways in which globalization has influenced the wine industry is through the concept of terroir, which refers to the unique combination of soil, climate, and topography that give wines their distinctive characteristics. As winemakers from different regions interact and collaborate, they are able to experiment with new terroirs and grape varieties, leading to the production of wines that reflect a blend of traditional practices and modern influences. This cross-pollination of ideas has enriched the diversity of wines available to consumers, shaping the global wine market and driving innovation in winemaking techniques. Globalization has also facilitated the dissemination of scientific research and technological innovations in the wine industry. As winemakers around the world seek to improve the quality and consistency of their wines, they are increasingly turning to research and scientific advancements to enhance their understanding of the chemical processes that govern wine production. By sharing knowledge and best practices on a global scale, the wine industry is able to push the boundaries of what is possible in winemaking, leading to the development of new techniques and approaches that elevate the overall quality of wines produced worldwide.

International Trade in the Wine Industry

When considering the international trade of wine, it is essential to acknowledge the significant impact it has on the global market. Regions like France, Italy, Spain, and the United States are among the top producers and exporters, playing a crucial role in shaping the industry. The trade agreements between these countries often dictate the flow of wine across borders, affecting prices, availability, and consumer preferences worldwide. The exchange of different styles and varieties of wine enriches the market, offering consumers a diverse selection to choose from. The international trade in the wine industry is not limited to finished products but also includes raw materials, equipment, and expertise. Winemaking technology and techniques are often shared between countries, leading to innovation and quality improvements in production processes. The transfer of knowledge between winemakers from different regions contributes to the continuous development and evolution of the industry, creating a dynamic and competitive environment. This collaboration fosters growth and sustainability in the market, benefitting both producers and consumers alike. With the rise of globalization, the demand for wine from emerging markets has increased significantly. Countries in Asia, South America, and Africa have shown a growing interest in wine consumption, prompting established wine-producing regions to adapt and cater to these new markets. The expansion of trade networks and distribution channels allows for greater accessibility to a diverse range of wines, opening up opportunities for producers to reach a wider audience. As the industry continues to evolve, international trade will play a crucial role in shaping the future of the wine market, driving innovation, diversity, and consumer experiences.

Cross-Cultural Influences on Wine Preferences

One of the key factors that influence wine preferences is the cultural background of consumers. Different cultures have unique tastes and traditions when it comes to wine, which can shape their preferences for certain types of wine. In some cultures, red wine is favored for its bold and robust flavors, while in others, white wine is preferred for its lighter and more delicate taste. These cultural influences can be traced back to historical and social factors, such as traditional winemaking practices, religious beliefs, and culinary customs. Understanding these cross-cultural influences is essential for wine producers and marketers looking to cater to diverse consumer preferences in a global market. The concept of terroir also plays a significant role in shaping wine preferences across different cultures. Terroir refers to the unique combination of soil, climate, and geography that gives each wine region its distinct characteristics. This sense of place can greatly impact the flavors and aromas of wine, leading to regional preferences for specific styles of wine. Wines from cool-climate regions may exhibit higher acidity and lighter body, appealing to consumers who enjoy crisp and refreshing wines. By considering the influence of terroir on wine profiles, producers can better tailor their products to meet the preferences of consumers in different cultural contexts. In addition to cultural and terroir influences, advancements in global communication and travel have also contributed to the cross-cultural exchange of wine preferences. As consumers become more exposed to a variety of wines from around the world, their tastes and preferences are likely to evolve and become more diverse. This increasing globalization of wine consumption presents both challenges and opportunities for producers, who must navigate the complex interplay of cultural norms, regional characteristics, and evolving consumer trends. By embracing this cross-cultural exchange, the wine industry can continue to innovate and adapt to meet the ever-changing demands of a multicultural market.

Challenges and Opportunities in Global Wine Market

One of the challenges in the global wine market is the increasing competition among wine producers from different regions around the world. With advancements in technology and transportation, consumers have access to a wide variety of wines, leading to intense competition for market share. This competition has put pressure on wineries to differentiate themselves through quality, branding, and marketing strategies. Fluctuations in consumer preferences and economic factors can create uncertainties for wine producers, making it challenging to predict market demands accurately. Despite these challenges, the global wine market also presents several opportunities for growth and innovation. The growing trend towards organic and sustainable winemaking practices has opened up new markets for environmentally-conscious consumers. By adopting these practices, wineries can differentiate themselves and appeal to a niche market segment. Emerging markets in countries such as China and India offer significant growth opportunities for wine producers looking to expand their reach. By understanding the unique preferences and trends in these markets, wineries can tailor their products to meet the demands of a diverse consumer base. The challenges and opportunities in the global wine market call for strategic approaches from wine producers to stay competitive and capitalize on growth prospects. By focusing on quality, innovation, and sustainability, wineries can navigate the complexities of the market and secure a strong position in the industry. With the right strategies in place, wine producers can continue to thrive in a dynamic and evolving market landscape, while meeting the diverse needs of consumers worldwide.

XXXI. WINE AND CLIMATE CHANGE

Over the past few decades, the wine industry has been facing the challenges posed by climate change. Rising global temperatures, unpredictable weather patterns, and shifting growing seasons are affecting vineyards worldwide. The impact of climate change on wine production is multifaceted, influencing not only grape yields but also the quality and characteristics of the wine produced. One of the key ways in which climate change is affecting the wine industry is through shifts in grape ripening times. Warmer temperatures are causing grapes to ripen earlier than usual, leading to changes in sugar, acidity, and phenolic compounds in the grapes. This can result in wines with higher alcohol levels, lower acidity, and altered flavor profiles. Winemakers are having to adapt their harvesting schedules and vineyard management practices to cope with these changes and maintain the desired quality of their wines. Climate change is also impacting the terroir of wine regions, influencing factors such as soil composition, water availability, and sunlight exposure. These changes can affect the overall character of wines, making it challenging for winemakers to maintain consistency in quality from year to year. As a result, the wine industry is exploring new techniques and technologies to mitigate the effects of climate change and ensure the sustainability of wine production for future generations.

Impact of Climate Change on Wine Production
Climate change poses a significant threat to wine production around the world. Rising global temperatures have already started to impact the grapes grown in many traditional wine

regions. Higher temperatures can lead to changes in grape ripening patterns, affecting the balance of sugars, acids, and phenolic compounds in the grapes. This, in turn, can alter the flavor profile and quality of the wines produced from these grapes. Changes in precipitation patterns can result in droughts or excessive rainfall, both of which can have negative consequences for vine growth and grape quality. One of the key ways in which climate change affects wine production is through shifts in terroir. The unique combination of soil, climate, and topography that defines a wine regions terroir plays a crucial role in determining the quality and characteristics of the wines produced there. As the climate changes, these factors may be altered, leading to changes in the expression of terroir in the wines. Warmer temperatures may lead to lower acidity levels in the grapes, while changes in rainfall patterns can affect soil moisture levels and nutrient availability, impacting grape growth and wine quality. In response to these challenges, many wine producers are implementing strategies to adapt to a changing climate. These may include changes in vineyard management practices, such as adjusting irrigation schedules or planting more drought-resistant grape varieties. Some wineries are also experimenting with new winemaking techniques to mitigate the effects of climate change on the final product. By understanding the complex interactions between climate, terroir, and grape chemistry, winemakers can make informed decisions to preserve the unique characteristics of their wines in the face of a changing climate.

Adaptation Strategies for Vineyards

One key aspect of winemaking that has gained increasing attention in recent years is the implementation of adaptation strategies for vineyards. Due to the changing climate and environmental conditions, vineyard managers and winemakers are faced with the challenge of ensuring the quality and quantity of grape production while maintaining sustainability. One such strategy involves the careful selection of grape varieties that are more resilient to changing climate patterns, such as drought-resistant cultivars that can thrive in arid conditions. By choosing the right grape varieties, vineyards can adapt to the challenges posed by climate change and ensure a consistent supply of grapes for winemaking. In addition to selecting appropriate grape varieties, vineyards can also utilize innovative irrigation techniques to optimize water usage and minimize water wastage. Methods such as drip irrigation and precision irrigation systems can help vineyards efficiently deliver water to the grapevines based on their specific needs, reducing water consumption and promoting healthier vine growth. Adopting sustainable farming practices, such as cover cropping, organic fertilization, and integrated pest management, can help vineyards build resilience against climate change impacts and improve the overall health of the vineyard ecosystem. Another important adaptation strategy for vineyards is the implementation of climate-smart viticulture techniques, which aim to mitigate the effects of climate change on grape production. These techniques may include canopy management practices, such as leaf thinning and shoot positioning, to optimize sunlight exposure and air circulation within the vine canopy, leading to improved grape ripening and quality. Vineyards can employ advanced weather monitoring technologies to track microclimatic conditions and make informed decisions regarding irrigation, pest control, and harvest timing. By integrating these adaptation strategies, vineyards can enhance their resilience to climate change and ensure the continued production of high-quality grapes for winemaking.

Influence of Temperature Variations on Grape Chemistry

Temperature variations play a crucial role in influencing the chemistry of grapes, which ultimately affects the quality and characteristics of wine. Fluctuations in temperature can impact the accumulation of key compounds in grapes, such as sugars, acids, and phenolic compounds. Warm temperatures can accelerate sugar accumulation in grapes, leading to higher alcohol levels in the resulting wine. On the other hand, cooler temperatures can help grapes retain higher acidity, which is essential for balancing the sweetness in wine. These variations in temperature can also influence the development of phenolic compounds, such as tannins, which contribute to the structure and mouthfeel of the wine. In addition to affecting the chemical composition of grapes, temperature fluctuations during the growing season can also impact the overall flavor profile of wine. Cooler temperatures can enhance the retention of aromatic compounds in grapes, leading to more pronounced and complex aromas in the finished wine. Conversely, excessive heat during ripening can cause aromatic compounds to degrade, resulting in wines with less expressive aromas. Understanding the influence of temperature on grape chemistry is essential for winemakers to make informed decisions regarding harvest timing, fermentation techniques, and overall wine quality. The influence of temperature variations on grape chemistry underscores the intricate relationship between viticulture practices and wine quality. By carefully monitoring and adjusting for temperature fluctuations throughout the growing season, winemakers can optimize the chemical composition of grapes to achieve desired flavor profiles in the final wine. The impact of temperature variations on grape chemistry highlights the importance of terroir in wine production, as climate and microclimate conditions play a significant role in shaping the unique characteristics of different wine regions. A deeper understanding of how temperature influences grape chemistry can help winemakers produce wines that truly reflect the terroir and showcase the best qualities of the grapes.

XXXII. WINE LABEL DESIGN AND MARKETING

One crucial aspect in the marketing of wine is the design of the label. Wine label design plays a significant role in attracting consumers and conveying important information about the product. The label serves as a visual representation of the brand and can influence a consumers perception of the quality and value of the wine. A well-designed label can help the product stand out on crowded shelves and create a lasting impression on potential buyers. Elements such as color, typography, imagery, and overall design aesthetic are carefully considered to appeal to the target market and convey the unique characteristics of the wine. In addition to aesthetic considerations, wine label design also serves a practical purpose by providing key information to consumers. Details such as the grape varietal, region of origin, vintage year, alcohol content, and producer name are typically included on the label. This information helps consumers make informed decisions when selecting a wine and provides context for understanding the products quality and characteristics. The label can also communicate the story and values of the winery, further enhancing the consumers connection to the product and brand. Effective wine label design strikes a balance between visual appeal and informative content, ultimately influencing consumer behavior and purchase decisions. Wine label design is closely tied to marketing strategies aimed at creating a strong brand identity and building customer loyalty. Consistency in label design across a wineries product line helps establish brand recognition and differentiate the products from competitors. By incorporating unique design elements and branding elements, wineries can build a cohesive brand image that resonates with consumers and sets them apart in a competitive market. Innovative and creative label designs have the potential to generate buzz and attract new customers, as well as engage existing ones. In today's digitally-driven marketplace, where consumers are bombarded with choices, a well-crafted label can make a significant impact on the success of a wine brand.

Importance of Label Design in Branding

Label design plays a crucial role in branding, as it serves as the visual representation of a wine producers identity and message. A well-designed label can convey the story behind the wine, the wineries values, and the quality of the product itself. Through the use of color, typography, imagery, and other design elements, a label can attract consumers, evoke emotions, and create a lasting impression. This visual aspect is often the first point of contact between the consumer and the wine, influencing the purchasing decision and building brand loyalty. Label design can differentiate a wine brand in a crowded market, helping it stand out on the shelf or online. By creating a unique and memorable label, wineries can establish a recognizable identity that sets them apart from competitors. This distinctiveness can be a powerful tool in capturing the attention of consumers and building brand awareness. A well-designed label can also communicate important information about the wine, such as the varietal, region, vintage, and tasting notes, helping consumers make informed choices when selecting a bottle. Label design can reflect the characteristics of the wine itself, providing cues to the sensory experience that consumers can expect. The choice of colors, images, and text on the label can evoke the flavor profile, aroma, and texture of the wine, creating anticipation and enhancing the overall drinking experience. A cohesive and harmonious design can align with the wines sensory qualities, creating a holistic and immersive brand experience for consumers. In this way, label design not only serves as a marketing tool but also as a way to enhance the storytelling and sensory journey of the wine.

Marketing Strategies for Different Wine Categories

When considering marketing strategies for different wine categories, it is crucial for producers and marketers to understand the varying preferences of consumers. When promoting red wines, it is essential to highlight their rich and robust flavors, as well as their potential health benefits. Marketing campaigns for white wines, on the other hand, can focus on their crisp acidity, fruit-forward profiles, and versatility when paired with different types of cuisine. Rosé wines may appeal to a younger demographic seeking a light and refreshing option, making social media platforms an effective marketing tool to target this audience. In addition to understanding consumer preferences, marketing strategies for different wine categories should also take into account the unique characteristics of each type of wine. For sparkling wines, emphasizing their celebratory nature and versatility for various occasions can be a successful approach. Marketing strategies for organic or sustainably produced wines should highlight their environmentally friendly practices and commitment to quality. By tailoring marketing campaigns to emphasize these distinguishing features, producers can effectively target niche markets and attract consumers seeking specific attributes in their wine selections. The use of storytelling and branding can play a significant role in marketing different wine categories. By creating compelling narratives around the wineries heritage, winemaking techniques, or the unique terroir of the vineyard, producers can establish an emotional connection with consumers. Leveraging social media platforms, influencer collaborations, and experiential marketing can also help engage consumers and create buzz around specific wine categories. A successful marketing strategy for different wine categories should combine an understanding of consumer preferences, a focus on highlighting unique characteristics, and the use of storytelling to create a memorable and compelling brand presence in the market.

Consumer Perception of Wine Labels

When considering the complex world of wine appreciation, one crucial yet often overlooked aspect is the role of wine labels. These labels are not merely decorative; they serve as powerful tools for conveying information and shaping consumer perception. Research has shown that wine labels play a significant role in influencing consumer decision-making, with studies indicating that consumers often make judgments about the quality and taste of wine based on the label design. Beyond aesthetics, wine labels also provide vital information about the wines origin, grape variety, vintage, and producer, helping consumers navigate the vast array of options available on the market. Wine labels can evoke emotions and associations that influence consumer preferences. A well-designed label that communicates authenticity, tradition, or sustainability can create a positive perception of the wine and its producer. Conversely, a poorly designed label that conveys a lack of attention to detail or quality may deter consumers from trying the wine, regardless of its actual quality. In this sense, wine labels serve as a form of communication between producers and consumers, shaping perceptions and expectations before the bottle is even opened. Understanding the psychological impact of wine labels on consumer perception is crucial for wineries seeking to differentiate themselves in a competitive market and build a loyal customer base. The design and content of wine labels are essential considerations for wineries aiming to capture the interest and loyalty of consumers. By carefully crafting labels that convey key information, evoke positive emotions, and align with consumer preferences, wineries can enhance the overall experience of wine consumption and differentiate their products from competitors. As the wine industry continues to evolve, with new trends and innovations shaping consumer preferences, the role of wine labels in influencing perception and purchase decisions will undoubtedly remain a critical aspect of marketing and branding strategies.

XXXIII. WINE AND SOCIAL DYNAMICS

One interesting aspect of wine that often goes beyond its chemical composition is its role in shaping social dynamics. Wine has been associated with social gatherings, celebrations, and rituals for centuries, playing a significant role in cultural and societal interactions. The sharing of a bottle of wine can create a sense of community and conviviality among individuals, fostering connections and enhancing social relationships. In many cultures, the act of sharing wine is also seen as a gesture of hospitality and friendship, strengthening bonds and promoting social cohesion. This social aspect of wine consumption highlights its importance beyond its taste and chemical components, emphasizing its role in shaping human interactions and relationships. The concept of wine appreciation and wine culture has evolved into a sophisticated social activity that involves a certain level of knowledge and expertise. Wine tasting events, wine tours, and wine clubs have become popular settings for enthusiasts to explore different varieties, regions, and vintages of wine. These experiences not only allow individuals to expand their palate and deepen their understanding of wine appreciation but also provide opportunities for social interaction and networking. The shared interest in wine can serve as a common ground for people to come together, bond over their passion for wine, and engage in meaningful conversations, thus enriching their social experiences and connections. Wine has the ability to transcend social boundaries and bring people from diverse backgrounds together. Whether it is a casual gathering among friends, a formal business dinner, or a cultural event, wine often serves as a unifying element that promotes communication and breaks down barriers. The act of sharing and enjoying wine can create a sense of camaraderie and mutual understanding among individuals, regardless of their differences. This demonstrates the powerful role that wine plays in facilitating social interactions, fostering connections, and promoting a sense of community among people from various walks of life. In essence, the social dynamics of wine go beyond its chemical makeup, illustrating its significance in shaping human relationships and interactions.

Wine Consumption Trends in Different Demographics

Current trends in wine consumption reveal interesting patterns among different demographics. One notable trend is the increasing popularity of wine among younger consumers, particularly Millennials and Generation Z. These younger generations are drawn to wine for its perceived health benefits, social appeal, and versatility in pairing with a variety of foods. The rise of wine influencers and digital platforms showcasing wine culture has contributed to this demographics interest in exploring different wine varietals and styles. As a result, wineries are adapting their marketing strategies to target these younger consumers, emphasizing sustainability, innovation, and authenticity in their products. Conversely, older demographics, such as Baby Boomers, are continuing to drive wine consumption, albeit in different ways. This group tends to prioritize traditional wine regions, established vineyards, and recognizable wine brands. They often have more disposable income to spend on premium wines and are more likely to attend wine tastings, tours, and events. Wineries catering to this demographic focus on heritage, legacy, and quality craftsmanship to appeal to their preferences.

Despite these generational differences, the overall trend indicates a growing interest in wine across various age groups, with each demographic influencing the market in unique ways. There are regional variations in wine consumption trends, influenced by factors such as cultural customs, economic conditions, and accessibility to different wine regions. In countries with a long history of winemaking, such as France, Italy, and Spain, wine is deeply embedded in the cultural fabric, leading to consistently high levels of wine consumption among the population. In contrast, emerging wine markets in regions like China, India, and South America are experiencing rapid growth in wine consumption as incomes rise and consumer preferences shift towards premium and imported wines. Understanding these demographic and regional trends is crucial for wineries and wine producers to adapt their products, marketing strategies, and distribution channels to meet the evolving demands of diverse consumer groups.

Social Influences on Wine Preferences

Social influences play a significant role in shaping individuals wine preferences. Cultural norms, social status, and peer influences all play a part in determining what types of wines individuals may prefer. In some cultures, certain wines are prized for their historical significance or symbolic value, leading individuals to develop a taste for these specific varieties. Individuals may gravitate towards wines that are associated with higher social status, as a way to signal their sophistication or wealth to others in their social circle. Peer influences can also impact wine preferences, as individuals may be influenced by the preferences of their friends or family members. Wine tasting events, social gatherings, and recommendations from peers can all shape an individual's perception of different wines and influence their preferences. This social aspect of wine consumption can play a crucial role in expanding individuals palates and introducing them to new varieties of wine that they may not have considered on their own. Social influences on wine preferences highlight the complex interplay between individual tastes and external factors. By understanding how cultural norms, social status, and peer influences can shape individuals wine preferences, wine producers and marketers can create targeted strategies to appeal to different segments of the market. Taking into account these social influences can lead to a more nuanced understanding of consumer behavior and help enhance the overall wine-drinking experience for individuals across diverse backgrounds.

Wine's Role in Social Gatherings and Events

Social gatherings and events have long been intertwined with the consumption of wine, serving as a symbol of celebration, camaraderie, and conviviality. In many cultures, the act of sharing a bottle of wine has deep-rooted symbolic significance, signifying trust, unity, and a sense of community among those partaking in the experience. The ritual of pouring wine and raising a toast has the power to bring people together, breaking down barriers and fostering connections that transcend mere social niceties. As such, wine plays a vital role in facilitating meaningful interactions and creating unforgettable memories in the context of social gatherings, whether it be a formal dinner party, a casual get-together, or a momentous celebration. Wines multifaceted nature extends beyond its role as a mere beverage, as it also serves as a cultural touchstone that reflects the unique traditions, customs, and values of a particular region or community. The choice of wine at an event can be a deliberate expression of identity or status, conveying messages about taste, sophistication, and discernment. The act of sharing and discussing wine can engender a sense of shared experience and mutual appreciation among participants, fostering a sense of connection and belonging that transcends individual differences. In this way, wine becomes a conduit for cultural exchange and understanding, bridging gaps and fostering a sense of unity among diverse groups of people. In the realm of social gatherings and events, wine emerges as a versatile and dynamic element that has the power to enhance the overall experience and leave a lasting impression on participants. From its ability to stimulate conversation and spark lively debates to its capacity to elevate the sensory pleasure of a meal or occasion, wine adds an element of sophistication and refinement to any gathering. The sheer diversity of wine varietals, styles, and flavor profiles provides ample opportunity for exploration and discovery, encouraging participants to expand their palates and deepen their appreciation for the complexities of this beloved beverage. Wines role in social gatherings and events transcends mere consumption, as it becomes a catalyst for fostering connections, nurturing relationships, and creating moments of joy and conviviality that are cherished long after the last drop has been poured.

CONCLUSION

The chemistry of wine is a fascinating field that intertwines art and science to create a complex beverage enjoyed by millions around the world. Through the intricate interplay of acids, sugars, tannins, and aromatic compounds, the taste, aroma, and mouthfeel of wine are meticulously crafted. The fermentation process, driven by yeasts and bacteria, further contributes to the rich chemical composition of wine, with each strain playing a crucial role in shaping the final product. These chemical components vary among different types of wines, whether red, white, rosé, or sparkling, highlighting the versatility and diversity of the winemaking process. The influence of terroir and winemaking techniques cannot be understated when considering the chemical makeup of wine. The unique characteristics of a particular wine can be attributed to the geographical location where the grapes are grown, as well as the methods employed during the winemaking process. These factors further underscore the importance of understanding the chemistry behind wine production and the impact it has on the final product. As the wine industry continues to evolve, scientific advancements play a pivotal role in shaping the future of winemaking, driving innovation and pushing the boundaries of what is possible in creating exceptional wines that captivate the senses. In summary, the chemistry of wine represents a harmonious blend of tradition, innovation, and scientific discovery. By exploring the intricate chemical components that define the taste, aroma, and texture of wine, we gain a deeper appreciation for the complexities involved in producing this beloved beverage. As the wine industry continues to evolve and embrace new technologies, the future holds great promise for further advancements in winemaking practices. Through a deeper understanding of the chemical processes at play, we can enhance our enjoyment and appreciation of the artistry behind every glass of wine.

BIBLIOGRAPHY

Percy H. Dougherty. 'The Geography of Wine.' Regions, Terroir and Techniques, Springer Science & Business Media, 1/3/2012

Helder Fraga. 'Viticulture and Winemaking under Climate Change.' MDPI, 12/19/2019

John Gladstones. 'Wine, Terroir and Climate Change.' Wakefield Press, 1/1/2011

Linda F. Bisson. 'Understanding Wine Microbiota: Challenges and Opportunities.' Aline Lonvaud, Frontiers Media SA, 8/16/2019

Charles G. Edwards. 'Wine Microbiology.' Practical Applications and Procedures, Kenneth C. Fugelsang, Springer Science & Business Media, 4/3/2007

Yves Glories. 'Handbook of Enology, Volume 2.' The Chemistry of Wine - Stabilization and Treatments, Pascal Ribéreau-Gayon, John Wiley & Sons, 5/1/2006

Steven N. Dalton. 'Recommended Packaging Specifications for BEER and BEVERAGE Bottles.' A guide for Beer and Beverage manufacturers, and Contract Filling companies, GPC Technical, 4/6/2021

Patrick Wiget. 'Wine Packaging. The Influence on the Consumer's Perception of Quality.' Bod Third Party Titles, 1/1/2020

Matthew Gedeon. 'Advanced Topics in Oenology.' Wine Packaging and Storage, CreateSpace Independent Publishing Platform, 1/1/2015

Iris Loira. 'Grape and Wine Biotechnology.' Antonio Morata, BoD – Books on Demand, 10/19/2016

Santosh Kumar. 'Nanotechnology Advancement in Agro-Food Industry.' Ragini Singh, Springer Nature, 8/24/2023

Sofia Catarino. 'Improving Sustainable Viticulture and Winemaking Practices.' J. Miguel Costa, Academic Press, 3/19/2022

Jean L. Jacobson. 'Introduction to Wine Laboratory Practices and Procedures.' Springer Science & Business Media, 6/14/2006

John F. Jackson. 'Wine Analysis.' Hans-Ferdinand Linskens, Springer Science & Business Media, 12/6/2012

Kenneth C. Fugelsang. 'Wine Analysis and Production.' Bruce Zoecklein, Springer Science & Business Media, 11/9/2013

Patrick Iland. 'Chemical Analysis of Grapes and Wine.' Techniques and Concepts, Patrick Iland Wine Promotions, 1/1/2004

Iris Loira. 'Yeast.' Industrial Applications, Antonio Morata, BoD – Books on Demand, 11/8/2017

John Hudelson. 'Wine Faults: Causes, Effects, Cures.' Board and Bench Publishing, 11/1/2010

Alina Maria Holban. 'Preservatives and Preservation Approaches in Beverages.' Volume 15: The Science of Beverages, Alexandru Grumezescu, Academic Press, 7/17/2019

T. Matthew Taylor. 'Antimicrobials in Food.' P. Michael Davidson, CRC Press, 11/10/2020

Fernanda Cosme. 'Grapes and Wines.' Advances in Production, Processing, Analysis and Valorization, António M. Jordão, BoD – Books on Demand, 2/28/2018

Charles Emil Calm. 'Sulphurous Acid and Sulphites as Food Preservatives.' Hygeian Chemical and Research Laboratory, 1/1/1904

Neville Vassallo. 'Polyphenols and Health.' New and Recent Advances, Nova Publishers, 1/1/2008

Cheryl Anne Combs. 'Tannins.' Biochemistry, Food Sources and Nutritional Properties, Nova Science Publishers, Incorporated, 1/1/2016

Alfredo Aires. 'Tannins.' Structural Properties, Biological Properties and Current Knowledge, BoD – Books on Demand, 1/8/2020

Sarah Troxell . 'Acidity Management in Musts & Wines.' Acidification, deacid-ification, and crystal stabilization, Volker Schneider, Board and Bench Publish-ing, 10/1/2018

Justin Hammack. 'Wine Folly.' The Essential Guide to Wine, Madeline Puckette, Penguin Publishing Group, 9/22/2015

Mary Griffin. 'The pH Scale.' Gareth Stevens Publishing LLLP, 12/15/2018

Jose Ramon Verde-Calvo. 'Flavour Science.' Chapter 50. Effect of Temperature during Bottle Aging on the Flavor Profile and Antioxidant Capacity of Ruby Cabernet Red Wine, Josefa Espitia-Lopez, Elsevier Inc. Chapters, 7/29/2013

H.G. Elias. 'Macromolecules · 1.' Volume 1: Structure and Properties, Springer Science & Business Media, 12/6/2012

Henry H. Work. 'Wood, Whiskey and Wine.' A History of Barrels, Reaktion Books, 8/12/2024

Donald Herbert Ford. 'Neurobiological Aspects of Maturation and Aging.' Else-vier, 1/1/1973

Andrzej Stankiewicz. 'Intensification of Biobased Processes.' Andrzej Gorak, Royal Society of Chemistry, 6/18/2018

Robert Henry Aders Plimmer. 'The Chemical Changes and Products Resulting from Fermentations.' Longmans, Green, 1/1/1903

Jamil Zainasheff. 'Yeast.' The Practical Guide to Beer Fermentation, Chris White, Brewers Publications, 2/1/2010

Brycen Soto. 'Fermentation Processes.' Scientific e-Resources, 3/20/2019

Antonio Morata. 'Red Wine Technology.' Academic Press, 10/29/2018

William B. Alwood. 'Enological Studies.' The Chemical Composition of American Grapes Grown in Ohio, New York, and Virginia, U.S. Government Printing Office, 1/1/1912

Julia Harding. 'Wine Grapes.' A Complete Guide to 1,368 Vine Varieties, Including Their Origins and Flavours, Jancis Robinson, Harper Collins, 9/24/2013

William Bradford Alwood. 'Enological Studies.' The Chemical Composition of American Grapes Grown in Ohio, New York, and Virginia, U.S. Department of Agriculture, Bureau of Chemistry, 1/1/1911

Narayan changder. 'Research methodology.' the amazing quiz book, changder outline, 12/21/2022

Gavin L. Sacks. 'Understanding Wine Chemistry.' Andrew L. Waterhouse, John Wiley & Sons, 5/16/2024

www.ingramcontent.com/pod-product-compliance
Lightning Source LLC
Chambersburg PA
CBHW070010300526
45794CB00001B/264